Alternative East Asian Nuclear Futures

Volume II: Energy Scenarios

AF879172

Alternative East Asian Nuclear Futures

Volume II: Energy Scenarios

Edited by

Henry D. Sokolski

Nonproliferation Policy Education Center

Copyright © 2018 by Henry D. Sokolski
Nonproliferation Policy Education Center
Arlington, VA 22209
www.npolicy.org

Printed in the United States of America

All rights reserved. Except for brief quotations in a review, this book, or parts thereof, must not be reproduced in any form without permission in writing from the Nonproliferation Policy Education Center.

ISBN 978-0-9862-895-6-9

Cover images, from left to right: 1) Image of the Fukushima Daiichi nuclear disaster on March 11, 2011 in Japan; 2) map of a potential North East Asia 'Supergrid'; and 3) a ship transporting liquefied natural gas (LNG).

CONTENTS

INTRODUCTION

Alternative East Asian Nuclear Futures

Henry D. Sokolski

The 13 chapters contained in this book's two volumes were prompted by a single inquiry in 2012 from the MacArthur Foundation. Was there any way, I was asked, to further clarify the economic and nonproliferation downsides if further production of civilian plutonium proceeded in East Asia? My initial reply was no. So much already had been done.

But the more I thought about it, two things that had yet to be attempted emerged. The first was any serious analysis of just how bad things could get militarily if Japan and South Korea acquired nuclear weapons and North Korea and Mainland China ramped up their own production of such arms. Such nuclear proliferation had long been assumed to be undesirable but nobody had specified how such proliferation might play out militarily. Second, no serious consideration had yet been given to how East Asia might be able to prosper economically *without* a massive buildup of civilian nuclear power. Since each of the key nations in East Asia—China, the Koreas, and Japan—all would likely exploit their civilian nuclear energy infrastructure to acquire their first bombs or to make more, such inattention seemed odd.

What followed was encouragement from foundation staff; development of a proposal; funding from The MacArthur Foundation, The Carnegie Corporation, the Scaife Foundation, and The Smith Richardson Foundation; and more than four years of work. First,

I commissioned the very best regional security experts I could find to develop scenarios for Japan and South Korea acquiring nuclear weapons and for North Korea and China significantly ramping up their production and development of nuclear arms. These scenarios specified what each of these countries might do if they acquired a nuclear arsenal or expanded their existing stockpiles over the next 20 years.

Second, the authors presented these studies to leading Chinese, Russian, South Korean, Japanese, and American security and energy officials and experts at a series of workshops. The aim of these meetings was to get the participants' views on how real or worrisome the military scenarios might be. All of the military nuclear projections exploited civilian nuclear infrastructure to make nuclear weapons. Finally, to balance these dark nuclear projections, I commissioned a number of energy experts from East Asia and the United States to evaluate peaceful alternative civilian energy futures for East Asia that would rely on less nuclear power through 2035.

Privately, I was told that the project was doomed. No one of interest, I was told, would agree to participate. If they did, they wouldn't say anything of interest. And if they did, the participants wouldn't get along. Just the opposite occurred. Senior officials from each country did come; they all were candid; and the gatherings were surprisingly collegial.

In 2012, the project's first premise—that Japan and South Korea might use their civilian nuclear infrastructure to acquire nuclear weapons and that North Korea and China would perfect much more robust nuclear forces of their own—seemed fantastic. The prevailing wisdom was that Japanese or South Korean acquisition of nuclear weapons was unthinkable. It was not in their interest. Severe trade sanctions would be imposed upon them for violating the Nuclear Nonproliferation Treaty (NPT). Worse, such proliferation would weaken essential security ties with the United States. As for China and North Korea, most experts believed that neither would need nor want many nuclear weapons.

The project's second premise—that East Asia could meet its energy and environmental requirements without a large number of new reactors was also considered unlikely. At the time, most experts were arguing just the opposite, that the economies of Japan, South Korea, Mainland China, and Taiwan would falter without a massive build out or restart of planned nuclear power plants.

Accepted as wisdom six years ago, today none of these views seem particularly persuasive.

Today, our East Asian Allies are increasingly interested in developing nuclear weapons options.[1] In response to North Korea's nuclear saber-rattling, more than a few former and current officials both in South Korea and Japan—including former defense ministers and the leaders of the ruling and opposition parties—have come out in support of acquiring nuclear arms or a nuclear weapons option. Mostly, their enthusiasm for nuclear options has been driven by fear. On the other hand, some of this bravado is reasoned: In specific, it is no longer clear, if it ever was, that South Korea or Japan would suffer economically if they withdrew from the NPT. Consider India: Since 2011 it's been able to enjoy all of the civilian nuclear trade privileges of a member state of the Nuclear Supplier Group (NSG) despite being a nuclear-armed non-NPT state. This suggests other countries can acquire nuclear weapons, be outside of the NPT, and still skirt nuclear trade sanctions as well. Why wouldn't Washington be as forgiving of Seoul and Tokyo as it has been of New Delhi? Did Israel's, the UK's or France's acquisition

---

1. See, e.g., Michelle Ye Hee Lee, "More than ever, South Koreans want their own nuclear weapons," The Washington Post, September 13, 2017, available from *https://www.washingtonpost.com/amphtml/news/worldviews/wp/2017/09/13/most-south-koreans-dont-think-the-north-will-start-a-war-but-they-still-want-their-own-nuclear-weapons/*; Jake Adelstein, "Is Japan about to Is Japan About to Hit Its Nuclear Tipping Point?" *The Daily Beast*, February 15, 2018, available from *https://www.thedailybeast.com/is-japan-about-to-hit-its-nuclear-tipping-point*; and David E. Sanger, Chloe Sang-Hun, and Motoko Rich, "North Korea Rouses Neighbors to Reconsider Nuclear Weapons," *The New York Times*, October 28, 2017, available from *https://www.nytimes.com/2017/10/28/world/asia/north-korea-nuclear-weapons-japan-south-korea.html*.

of nuclear arms terminate security ties with Washington? Presumably, officials both in Seoul and Tokyo know the answers as well as those in Washington. If not, they need only reflect on the North Korean case: It withdrew from the NPT in 2002 and suffered no specific sanctions at all.

As for the weapons ambitions of China and North Korea, they too no longer look to be so limited. China, faced with both Russian and American nuclear arsenal revitalization programs and a perceived increased willingness to threaten use, has announced that it will need to increase and upgrade its stockpile as well.[2] North Korea, meanwhile, has shown no restraint at all. It not only seems intent on increasing the number of nuclear weapons in its arsenal (now projected to grow to more than 100 by 2030), but to test and deploy sea and ground-based missiles of nearly all sorts. It remains to be seen how what they've built might be used as leverage for political and economic concessions during negotiations with the United States.

This, then, brings us to the further growth of nuclear power in East Asia. Today, nuclear power's expansion in Asia is in retreat. Taiwan plans to go nonnuclear by 2025; South Korea by 2030. The Japanese government is eager to restart as many as possible of the 54 reactors it had online before the Fukushima accident shut them down. As of this writing, however, Japan has only eight online and many are slated to be shuttered.[3] The big question is whether and to what extent Japan will further adapt its electrical system to allow nonnuclear alternatives a greater role in the country's electrical power mix. Finally, Mainland China, once projected to have 200 gigawatts of electrical capacity on line by 2030, is encountering difficulties

2. See "China must expand nuclear arsenal in response to Trump: Global Times," *Strait Times*, January 24, 2017, available from *https://www.straitstimes.com/asia/east-asia/china-must-expand-nuclear-arsenal-in-response-to-trump-global-times*.

3. See Reuters Staff, "Table-Japan nuclear reactor operations: Kyushu shuts Sendai No. 2," *Reuters*, April 23, 2018, available from *https://af.reuters.com/article/commoditiesNews/idAFL3N1S02UE*.

and now may be lucky to have a bit less than 100 gigawatts on line by then.[4]

Nor have economies in East Asia suffered significantly because of this nuclear slowdown. China's economy continues to grow, albeit at a somewhat slower rate as its economy matures. Japan, South Korea, and Taiwan all have sustained positive growth since 2012. Meanwhile, increases in electrical demand in each of these economies have slowed. Certainly, projections that East Asian economies would tank without a ramp up in nuclear power have yet to materialize. Meanwhile, the price of liquefied natural gas, which has been used to help substitute for nuclear power, has fallen roughly 60% since 2012.

Much of this volume's commissioned research predicted these trends. These chapters are worth reading if only to understand what premises the authors used to reach their conclusions. North Korea is interested not in deterring its adversaries, but in coercing them and in driving U.S. forces out of the region. The outcome of its negotiations with the United States remains to be seen but it appears that Pyongyang will do so as a nuclear-armed equal. China, meanwhile, may have no choice but to expand its nuclear arsenal as it develops modern forces that demand the integration of warheads with missile delivery systems (submarines and road-mobile missiles) that are increasingly autonomous. Beijing also must deal eventually with a Russia that it will have difficulty trusting, or at least a Russia that is unwilling to trust Beijing.

Natural gas prices increasingly will be less local and instead reflect global market trends, which will be almost entirely independent of oil prices. Such globalized natural gas is likely to remain plentiful and affordable, at least for several decades.[5] Nuclear power from

4. See Steve Kidd, "Nuclear in China—why the slowdown?" *Nuclear Engineering International*, August 10, 2017, available from *http://www.neimagazine.com/opinion/opinionnuclear-in-china-why-the-slowdown-5896525/*.

5. For a thorough update on the likely glut of natural gas in East Asia, see Melanie

large reactors and nuclear recycling costs and construction times are high and unlikely to fall. By contrast, electrical demand growth, the cost of renewables, natural gas-fired electricity, and electrical grid storage are falling and are likely to continue to do so. Meanwhile, improving grid transmission systems and developing more agile and experimental electrical pricing systems can help to reduce demand and encourage more economical forms of electrical generation. All of these developments corroborate the key finding in most of the chapters that follow.

Some of the volume's research, though, has yet to be vindicated. The encouragement of market-driven competitions among energy types in the production of electricity has only begun in Japan, China, and South Korea on a small scale. It is likely to change over the next two decades. Nor has there been any significant commerce in electricity between East Asian states. Both developments, if they should occur, however, could dramatically increase the supply of electrical power and reduce requirements for ever more generating stations.

What does this research then suggest to keep East Asia more peaceful? Three things. First, whatever the merits of nuclear power's expansion might be in East Asia, deferring the commercial use of plutonium-based fuels and the further expansion of uranium enrichment capacity makes both security and economic sense. Neither of these activities have any positive return on investment and increase the technical ability for China, Korea, and Japan to either ramp up their existing nuclear numbers or breakout to build an ever large batch of bombs.

Second, energy pricing, investment, and regulatory reforms in China, Taiwan, South Korea, and Japan that rely more on market signals than on central planning would help determine the appropriate level of nuclear power needed. The uncertainties regarding what the op-

---

Hart, Luke Bassett, and Blaine Johnson, "Do Not Fall for the Hype on U.S.-China Natural Gas Trade," Center for American Progress, April 18, 2018, available from *https://www.americanprogress.org/issues/security/reports/2018/04/18/449807/not-fall-hype-u-s-china-natural-gas-trade/*.

timal types and mixes of new and existing forms of electrical generation, storage, and distribution systems might be as significant. How, if at all, these new systems might relate to the transport sector and industrial and commercial heating and cooling markets is also unclear and will remain so for the next 20 to 40 years. Trying to pick which technologies will be the clear winners without heavy reliance on market signals is a prescription for regret.

Third, tying nonproliferation to efforts to strengthen American-Japanese and American-Korean relations is required now more than before. The need to pursue serious long-term planning toward this end cannot be emphasized enough. Such planning needs to go beyond conventional and nuclear military war gaming to deter, defend, and counter possible North Korean provocations. The United States and its allies need to work with Seoul to understand what might reduce its "need" or incentives to go nuclear. These drivers of South Korean interest in going nuclear are not just military in character. They also are social, historical, economic, and diplomatic and can be identified and mitigated. Certainly, reacting to their expression as crises may unfold is far less leveraged.

My center has already made efforts to act on all three recommendations. First, as part of a follow-on grant by the MacArthur Foundation, the Nonproliferation Policy Education Center (NPEC) made a number of trips to visit with senior officials in Seoul, Tokyo, and Beijing and explored the idea of these countries adopting a policy of deferring their plans to recycle plutonium-based fuels commercially. As a result of these exchanges, I worked with members of the State Department's International Security Advisory Board to get all of its members to back a U.S.-led initiative to encourage a commercial plutonium pause in East Asia. Unfortunately, this board's unanimous support came late in the Obama Administration's second term. Action was not taken.

With the election in 2016, President Donald Trump appointed Rex Tillerson as Secretary of State and he was briefed on the desirability of pushing a commercial plutonium pause in East Asia. Mr.

Mike Pompeo has just been installed as Tillerson's replacement. It is unclear if he has yet been briefed on the idea but it is clear that if the stated goal of North Korean denuclearization is to be achieved, it may require nuclear restraint on both the plutonium recycling and uranium enrichment not just in North Korea but in neighboring states as well. Time will tell if this idea is given a chance or not.

Second, NPEC is currently conducting a two-year project to assess nonnuclear alternatives to the further expansion of nuclear power systems both in the Middle East and in Mainland China and Taiwan. The key here is to compare the costs of different energy systems both economically and environmentally. The initial research (see David Von Hipppel's Appendix in Vol. II) suggests that China's nuclear power program, which is growing faster than any other nation's, may no longer be increasing quite so quickly. Nor is it clear that it will grow anywhere nearly as large as originally planned. At the moment it is unclear if South Korea's plans to go nonnuclear by 2030 and Taiwan's plans to do so by 2025 will be the model for the region or if nuclear power will continue to expand. Much will depend on how things unfold on the Mainland.

NPEC's most recent studies may help. They are aimed at encouraging the U.S. government to do its own more detailed analyses of its own as called for by Title V of the Nuclear Nonproliferation Act of 1978. They also are being shared with key officials in East Asia.

Finally, NPEC has begun collaborating with the National Defense University's Center for the Study of Weapons of Mass Destruction and its Program for Emerging Leaders to set up a long-term analysis and planning program in support of the U.S. government's current policy of preventing South Korea and Japan from acquiring nuclear weapons. This program has just begun. It remains to be seen if the project's products will attain policy traction.

Chapter 1

East Asian Energy Policies: Foreign and Military Implications

John Lee

In economic and military terms, East and Southeast Asia is the most dynamic region in the world. Since the Global Financial Crisis that began with the Lehman Brothers collapse in 2008, the region has been behind two-thirds of global economic growth. Asia's share of energy consumption is forecast to grow from about 35% currently to 47% in 2035 according to International Energy Agency (IEA) figures.[1] For the first time in history, Asia's collective military spending has overtaken Europe's, and growth in military spending is the most rapid of any region in the world.[2]

This has given rise to an expanding literature about imminent or likely "resource wars" between East Asia's great powers, especially over energy resources. Such arguments are based on the notion that scarcity will drive competition and exacerbate tensions in a region where military spending is rising and strategic competition is intensifying. This often leads to the conclusion that the more self-sufficient East Asian countries are when it comes to energy, the more secure they will become—possibly reducing regional strategic and military competition.

1. International Energy Agency, *World Energy Outlook 2016*, Paris: Organization for Economic Cooperation and Development/IEA, 2013.

2. International Institute for Strategic Studies, *The Military Balance 2013*, London: IISS, 2013.

These assumptions and arguments should not be taken for granted but put under greater scrutiny. Using China as a primary case study and also as a central player in terms of how its energy policy will determine strategic and military relations in East Asia, the paper will make the following arguments.

First, energy security as a stand-alone issue is important but not a dominant driver and shaper of current strategic affairs and military posture, and China's strategic and military posture in particular. Instead, energy security feeds into broader geostrategic and structural tensions in the region.

Second, enhancing energy security in the form of greater self-sufficiency for all countries will not necessarily lead to a more benign and stable strategic and military environment. On the contrary, reliance on open, efficient, and impartial commodity markets is the most effective approach in meeting one's energy needs. Attempting to own or lock-in a foreign energy asset is far less efficient and ill-suited to meeting regional demand. The risk that needs to be managed is ill-considered energy policies and developments in the future that might deepen the energy insecurity of East Asian countries, encourage the militarization of energy-related policies, and worsen pre-existing geostrategic and political tensions.

Third, encouraging market based energy interdependence and reliance on sea-borne trade might even have a dampening effect on strategic and military competition. This is so since it raises the domestic economic cost of regional instability, thereby increasing the incentives for strategic restraint if not cooperation.

The paper will then conclude with some suggested principles for energy policy and futures in East Asia, and the strategic benefits of market-based rather than securitized (or even militarized) self-help energy security approaches.

China as the Central Variable in East Asian Strategic and Energy Futures

China-centric analyses will often oversimplify the complexity of relations in East and Southeast Asia. Moreover, the energy (and related strategic policies of other East Asian countries) will also shape the future strategic and military landscape of the region, and influence China's responses in turn. However, focusing on China's rise in shaping the future strategic environment in East Asia is appropriate for a number of reasons.

The first reason is based on economics and development, which will drive future regional growth in energy consumption. While Asia will account for a projected 47% of global energy demand in 2035 according to IEA estimates, China could well account for almost half of this at 18% of global energy usage—rising from around 14% currently.[3] Up to 2030, China will alone account for an estimated 25-40% of the increase in world energy usage. By 2030, China will match American energy demand even if energy usage per person remains significantly lower.

In contrast, the forecast for energy usage by other East Asian powers in Japan and South Korea will decline in relative terms. For Japan, the decline in energy demand as a proportion of global demand will fall from about 15% currently to 9% in 2030. Over the same period, South Korea's will decline from about 7% currently to 6%.

The strategic ramifications of energy use are generally caused by rapid changes in demand and/or supply. Leaving aside the supply question, it is clear that projected increases in Chinese demand will be the greatest driver of change and therefore uncertainty in terms of energy security in the region. In other words, if significant increases in energy demand (or significant disruption to energy supply) is a major driver of potential instability and tension, then China's economic rise is the primary regional factor in that equation.

3. All figures are from the *World Energy Outlook 2013*.

Second, in examining any correlation between energy security and strategic/military posture, it is noteworthy that China dominates defense spending in the region. Of the $20.96 billion in Asian spending increases between 2011 and 2012, 62.4% was accounted for by East Asia, followed by Southeast Asia (13.8%), South Asia (12.4%), and Australasia (11.4%). Within East Asia, China was behind 50.9% of defense spending, Japan 29.5%, and South Korea 14.4%.[4]

When Asia (which includes South Asia) is taken as a whole, China heads the list at 32.5% of spending, Japan second at 18.9%, and South Korea third at 9.2%. If one considers that the most powerful Southeast Asian countries such as Singapore, Indonesia, and Thailand accounted for 3.1%, 2.5%, and 1.7% of regional spending respectively, Chinese military dominance over the region in budgetary terms is clear.

Moreover, when one considers that Chinese spending on the People's Liberation Army (PLA) has been growing at rates exceeding gross domestic product (GDP) growth over the past decade at over 15% growth per annum since 2001, and is projected to continue this trend in the years ahead, Chinese dominance in military spending will be even more pronounced. This is particularly true since Japan—with an almost stagnant economy—is unlikely to increase defense spending beyond 1.5% of GDP for the foreseeable future. Indeed, China is expected to match American defense spending in dollar terms by around 2030, based on current projections.

Third, the most significant change in the regional military balance is the dramatic increase in Chinese capabilities over the past two decades—even if these capabilities remain untested.

Of high relevance is that the PLA's rapidly improving capabilities in so-called anti-access/area denial technologies (based on submarines, ballistic missile technology, and cyber and other net-worked disruption enhancements) threaten to deny American and Japanese forces capacity for sea-control over the so-called First Island Chain

4. All figures are from *The Military Balance 2013*.

which surrounds China's maritime periphery and stretches from the Kuril Islands in the Russian far east, to Japan, northern Philippines, Borneo and Malaysia—something which the U.S. Seventh Fleet has enjoyed for over five decades.[5]

The possible change in the military balance—even if one assumes that the PLA Navy cannot exercise sea control over the First Island Chain into the future—has a number of ramifications. One is growing uncertainty as to the American willingness to suffer significant military costs as a result of any conflict with China (e.g., over Taiwan or in defending its Japanese ally in a battle over control of the disputed Senkaku/Diaoyu Islands)—casting doubt upon the reliability of the American capacity and/or willingness to protect the maritime commons and interests of its Asian allies.

These fears are in turn encouraging capitals such as Tokyo to enhance their capabilities (both stand-alone and inter-operable with the Seventh Fleet) as a form of balance against China—particularly in ballistic missile technology, intelligence, surveillance, and reconnaissance capacity, and cyber capacity.

In the past, uncontested American military and naval supremacy as well as allied reliance on its unrivaled security guarantees have kept East Asian rivalries suppressed. But any future transition away from security "free-riding" towards a "self-help" system of military balancing has unpredictable consequences. This is due to the historical rivalries that have been largely suppressed during the era of American primacy but not entirely eliminated. In East Asia, historical animosity and rivalry between Tokyo and Seoul remains, as do territorial disputes over the Takeshima/Dokdo/Liancourt islets between these two countries. Although both countries identify

---

5. See, Dean Cheng, "Countering China's A2/AD Challenge," *The National Interest*, September 20, 2013, available from *http://nationalinterest.org/commentary/countering-china%E2%80%99s-a2-ad-challenge-9099* and James Holmes, "China's Selective Access-Denial Strategy," *The National Interest*, December 3, 2013, available from *http://nationalinterest.org/commentary/chinas-selective-access-denial-strategy-9482*.

nuclear-armed North Korea as a common threat, the South Korean response to a rearming Japan—exacerbated by declining American military dominance—would be unpredictable.

All of this is to simply emphasize that analysis of the foreign and military implications of energy policies in East Asia cannot fail to place what China chooses to do, and how regional powers respond, at the center of any inquiry. In contrast, the energy policies of Japan and South Korea are not likely to change significantly over time, unless Tokyo and/or Seoul are reacting to adverse developments in Beijing's energy policy. Far more dependent on energy imports than China, Japan and South Korea are nevertheless highly comfortable with market-based approaches to meeting their energy needs that has depended on American strategic and military pre-eminence in Asia as the guarantor of peace and stability.

Geostrategic Mindset Versus Economic Pragmatism

China's future energy policies and strategic preferences are somewhat less predictable. Indeed, the point about the centrality of China in forecasting possible strategic and military futures vis-à-vis changes and developments in energy policy is further emphasized by the prevalence of a highly political mindset when it comes to the organization of the Chinese political-economy, especially with respect to the energy sector.

To be sure, Beijing is not alone in viewing access to energy imports as an inherent component of national interest. But in China, the definition of "energy security" is much stricter than in other energy-importing countries such as Japan or South Korea since Beijing considers not just reliable and uninterrupted access as critical, but also cheap supply of energy as essential to its national and domestic political interest. Moreover, while securing cheap and reliable access to foreign oil is seen as essential for mitigating economic risk in all oil-importing countries, securing such access is also essential for mitigating risks to the survival of the Chinese Communist Party's

(CCP) hold on power. In other words, the politicization of energy security in China occurs in a manner that does not apply to the same extent in oil-importing East Asian democracies.

Conflating economic risk and risk to the regime in China stems from the fact that the modern CCP largely stakes its legitimacy on the capacity to deliver rapid economic growth. A significant disruption to China's oil supply or a jump in prices is likely to lead to the twin forces of mass discontent: A stagnating economy and inflation caused by spikes in domestic energy prices. At a minimum, rapid growth is required to generate sufficient jobs and sustainability of incomes; it is not lost on modern Chinese authorities that double-digit inflation was one of the major reasons behind countrywide protests in 1989.

This link between energy security and maintaining rapid growth (and therefore regime security) has deepened due to the evolving drivers of China's growth in place since the mid-1990s. From this period onwards, fixed asset investment (and exports) replaced domestic consumption as the driver of economic growth.

Indeed, fixed-asset investment was behind around 40% of Chinese growth at the turn of this century, rising to current levels of 50-60%. During the global financial crisis (2008-2010), it drove over three-quarters of GDP growth. At current levels, the contribution made by fixed-asset investment is the highest of any major economy in recorded history.[6] Fixed-asset investment is an immensely energy-intensive form of economic activity, especially in an economy that uses energy extremely inefficiently compared to advanced industrialized peers.[7] Examining Chinese oil consumption over the last two decades makes this clear. From 1993-2010, oil consumption

6. See David Li, "Large Domestic Non-Intermediated Investments and Government Liabilities," working paper, Tsinghua University's Center for China in the World Economy, Beijing, April 4, 2006.

7. See Yuan Hu, "Energy conservation assessment of fixed-asset investment projects: An attempt to improve energy efficiency in China," *Energy Policy* 43, April 2012, pp. 327-334.

increased from 140 million metric tons to about 440 million metric tons.[8]

China's gradual realization that access to foreign oil was becoming an issue of utmost importance must be understood alongside the evolution of the Chinese political-economy from being largely household-driven driven in the first 10 years of reform from 1979-1989 to state-sector driven from the mid-1990s onwards—the opposite of what occurred in industrializing neighbors such as Japan, South Korea, and Taiwan. This means that the state-owned sector is disproportionately driving energy consumption and growth in demand in the economy. For example, in the first five months of 2011, it was estimated that six energy intensive sectors—electricity, steel, building materials, mining, chemical engineering, and petroleum which are all state-dominated—accounted for 43% of China's power consumption growth in 2011.[9]

Moreover, it is not surprising that in this state-dominated political-economy, energy is explicitly designated a "strategic sector," meaning that political officers and bodies are to exercise extensive control and oversight over all aspects of that sector. This is important because it is a structure that tends to conflate political with commercial interest more so than any other East Asian power (with the exception of North Korea).

This is clear from closer examination of the energy sector in China. In the first years after oil self-sufficiency ended in 1993, Beijing reorganized its oil (and gas) assets into two state-owned firms: The China National Petroleum Corporation (CNPC) and the China Petroleum and Chemical Corporation (Sinopec). CNPC is the dominant upstream player in the sector, and along with its listed entity, PetroChina, accounts for over 66% of China's oil output. Sinopec

8. *BP Statistical Review of World Energy 2011*, London: BP plc, June 2011.

9. "Energy-intensive industries account for 43% of China's power consumption growth," *Xinhua*, July 8, 2011, available from *http://www.chinadaily.com.cn/business/2011-07/09/content_12869411.htm*.

accounts for at least half of the country's downstream activities such as refining and distribution. The state-owned China National Offshore Oil Corporation (CNOOC) is close to being a monopolistic player in offshore oil exploration and production, with other state-owned giants such as Sinochem Group becoming more prominent in offshore oil distribution.

This is a contrast to Japan's domestic economy where the electricity industry, for instance, is dominated by 10 private firms, albeit in a vertically integrated structure and each enjoying virtual regional monopolies. Generating around 85% of the country's power, the balance is generated by the Electric Power Development Co. and Japan Atomic Power Company (both formerly state-owned but now privatized.)

Although Chinese state-owned enterprises (SOEs) are not mere puppets of the CCP and generally behave as normal commercial entities, the authority of political officials over these SOEs goes beyond what occurs in industrialized East Asian economies such as Japan and South Korea where many oil and gas SOEs have been privatized, and remaining SOEs almost fully commercialized.[10]

In China, the shares, and therefore assets, of SOEs are held by the State Assets Supervision and Administrative Commission (SA-

10. It is granted that to compete against Chinese oil and gas SOEs which receive generous lines of credit and other benefits from the CCP, Tokyo in particular is increasing government assistance to Japanese firms. For example, the Japanese government established Japan Oil, Gas and Metals National Corporation (JOGMEC), formerly the Japan National Oil Corporation, is tasked with ensuring stable supply of oil and gas to the Japanese market. It does this through JOGMEC administering subsidies to private Japanese oil and gas companies seeking upstream acquisitions offshore. However, unlike in China, Tokyo seeks to ensure energy security through public-private partnerships where private firms take the lead, in addition to a deregulation agenda domestically to improve efficiency and competitiveness. Moreover, political officials in Japan do not oversee major personnel appointments in Japanese energy companies or take the lead in defining objectives and targets for Japanese energy firms operating domestically or abroad. See, Petroleum Association of Japan, *Petroleum Industry in Japan 2012*, September 2012.

SAC), which in turn is controlled by and answerable to China's top administrative and legislative body, the State Council of the National People's Congress. The higher strategic objectives and purpose of SOEs are set by the leadership of the CCP. The vast, opaque and complicated system means that it is not always easy to trace the chain of decision-making throughout the country's political economy. But it is clear that the senior managers of all central SOEs are almost all senior members of the CCP.[11] The three most senior corporate positions (Party Secretary, Chairman, and CEO) of the largest centrally managed SOEs which include the energy companies are all appointed by the CCP's Central Organization Department (COD), after review and approval by the all-powerful Standing Committee of the Politburo. Almost all appointees are CCP members, and in many cases the CEO and Party Secretary is the same person. The appointment of all remaining senior executives is carried out by the SASAC, which consults with the COD.

By ensuring that only state-owned entities become the dominant and influential Chinese players in upstream and downstream domestic and international markets, the ability of powerful entities and individuals within the CCP to shape and execute energy policy is immeasurably enhanced. Since private-sector and independent entities are prevented from playing major roles in the energy sectors, commercial decisions tend to be disproportionately influenced by political interests and considerations—whether they are shaped by individual, Party, or national interests.

A mindset of maximizing control over all aspects of energy production, supply, and distribution is one thing. But if Beijing is hardwired to adopt a far more politicized mindset when it comes to thinking about domestic and international energy policy, it is also clear that China is nevertheless forced to pragmatically rely on the operation of free-markets when it comes to actually securing reliable and cheap access to foreign energy resources. This stems from

11. See Minxin Pei, *China's Trapped Transition*, Cambridge, MA: Harvard University Press, 2006.

the fact that the CCP has no choice but to do so.

Figures for 2010 reveal that 23% of China's offshore equity oil production was in Kazakhstan, 15% in both Sudan and Venezuela, 14% in Angola, 5% in Syria, 4% in Russia, and 3% in Tunisia. Nigeria, Indonesia, Peru, Ecuador, Oman, Colombia, Canada, Yemen, Cameroon, Gabon, Iraq, Azerbaijan, and Uzbekistan make up the remaining 20%.[12] Chinese offshore equity production amounts to around 28% of total current Chinese importing requirements, which were 4.8 million barrels per day in 2010.[13] National oil company (NOC) owned/controlled offshore sites are currently producing around 1.37 million barrels per day, and known new purchases of offshore sites suggest that Chinese NOCs' overseas equity production will reach around 2 million barrels per day by 2020,[14] which is significantly less than the official 2020 target of 4 million.[15]

For Beijing, diversifying its guaranteed sources of oil around the world is an essential hedge against disruption in normal offshore supply caused by commodity markets. The economic approach of heavy reliance on purchasing oil in international commodity markets leaves the Chinese economy exposed to spikes in oil prices. Such price spikes could be the result of political unrest in a major oil exporting country such as Saudi Arabia, geopolitical events that might lead to Western sanctions against a major supplier such as Iran, or a rise in global demand, which occurred just prior to 2008.

Hedging against the whims and vagaries of commodity markets means that Chinese NOCs participate in global commodity markets when conditions are benign. Although Chinese companies are not

---

12. Julie Jiang and Jonathan Sinton, *Overseas Investments by Chinese National Oil Companies*, Paris: International Energy Agency, February 2011, available from *www.iea.org/publications/freepublications/publication/overseas_china.pdf.*

13. U.S. Energy Information Administration, "Country Analysis Briefs–China."

14. Ibid.

15. Ibid.

transparent about transactions in oil markets, there is strong anecdotal evidence from 2008-2010 that Chinese NOCs sold a significant portion of their offshore equity oil on local and international markets under benign conditions instead of shipping the resource back into China.[16] This makes sense given the high cost of transporting oil from distant fields. Also, China does not have the domestic refining capabilities necessary to handle such additional volume, meaning it would have to rely on costly third-party refineries. This would be much more expensive than sourcing oil on international markets. But locking up resources through offshore equity oil gives China the option of hedging, or of bypassing commodity markets should they deteriorate.

The important point to be made here is that China is deeply conflicted when it comes to energy policy. On the one hand, the prevailing view in Beijing is that energy is "too important to be left to market forces alone," as one expert on Chinese energy policy puts it.[17] But its ownership of energy assets offshore will not be sufficient to meet its needs while procuring oil from outside commodity markets is inefficient and expensive.

The bottom line is that Beijing and the CCP will continue to feel deeply uncomfortable in relying on open markets to secure its energy and market-forces to determine pricing, and will persist in hedging against disruptions to supply and price spikes—without knowing whether its offshore hedging strategy will actually adequately shield its economy from such potential disruptions and price increases. This suggests that Beijing is reliant on regional and global commodity markets, but not necessarily committed to these.

But other countries can shape China's energy policies, even if they cannot change its distrustful mindset. So long as open market systems remain non-discriminatory, blind to strategic rivalries, and

16. Jiang and Sinton, p. 17.

17. Heinrich Kreft, "China's quest for energy," *Policy Review*, no. 139, October/November 2006.

continue to meet Chinese needs, Beijing will have little reason to deviate from pragmatic participation and support for such open systems, despite the persistent geo-strategic energy security mindset. This means that in energy policy, at least, one has the prospect of encouraging and persuading Beijing to behave as a "responsible stakeholder" within a liberal economic order.

Energy Security and China's Military Build-up

If Beijing's mindset towards energy security is inherently politicized and geostrategic in nature, is the country's strategic and military policy increasingly shaped by energy security concerns as proponents of future East Asian "resource wars" tend to assume?

Even though China can meet 90% of its energy needs from domestic sources, it continues to import around half of its oil. With oil still making up about 20% of its energy mix, it is projected that over 80% of the country's oil will be imported from foreign sources, rising from about 50% currently.[18] Even if China increases its use of coal, gas, and nuclear in its energy mix, oil will remain the dominant fuel for commercial and consumer transportation.

Moreover, almost all of China's oil imports now and into the future are received from oil tankers (currently about 80-85% foreign flagged)[19] passing through the Malacca Straits—with oil from pipelines from Russia and Central Asia unable to significantly meet China's needs. Even if the proposed pipeline through Myanmar, which begins in the Bay of Bengal and runs into Yunnan province, is completed, this has the capacity to import about 440,000 barrels

---

18. All figures are from the *World Energy Outlook 2013*.

19. See "China's Oil Security: Diplomacy, Economics and the prospects for Peaceful Growth," *StratRisks*, July 25, 2013 and Gabriel B. Collins and Andrew S. Erikson, "Chinese Efforts To Create A National Tanker Fleet," in Gabriel B. Collins, Andrew S. Erikson, Lyle J. Goldstein and William S. Murray, eds., *China's Energy Strategy: The Impact on Beijing's Maritime Policies*, Annapolis, MD: Naval Institute Press, 2008, pp. 84-85.

per day (bpd). A pipeline from Siberia can pump about 620,000 bpd into China's northern regions. A proposed pipeline from Kazakhstan has a likely capacity of 400,000 bpd.[20] Even if all goes well with these alternative oil routes, China currently imports around 5 million bpd, meaning that the majority of its oil imports will still be from tankers passing through the Malacca Straits. Besides, Chinese reliance on imported oil will grow to an estimated 13 million bpd by 2030. Given likely declining oil production from Russia and Central Asia in 2030, the overwhelming majority of this will be shipped from the Middle East.

This has given rise to a common perception that China's military build-up is primarily or largely the result of its energy insecurity.[21] In 2003, and recognizing China's increasing reliance on oil tankers through sea-lines-of-communication (SLOCs), China's then-President Hu Jintao spoke ominously about China's worsening "Malacca Dilemma." Although the Malacca Straits is a body of water stretching 1000km, parts of the Straits are as narrow as 15km wide. Those emphasizing absolute security (i.e., control) over security of supply fear a hostile power (namely the United States) controlling the Straits and blocking nearly all of its energy imports in order to cripple its economy—as occurred against the Japanese prior to and during World War Two.

Subsequently, and in addition to the seabeds in China's periphery being rich in hydrocarbons, it is widely assumed that China's naval build-up and maritime claims in the East and South China Seas are primarily about alleviating its own energy insecurities through eventually enjoying "command of the commons" in critical maritime zones. Short of sea control, some Chinese strategists emphasize the fact that China cannot even protect its seaborne energy sup-

20. "Weber Weekly Tanker Report—Week 25", Greenwich, CT: Charles R. Weber Company, June 21, 2013.

21. See, for example, Robert Kaplan, *Monsoon: The Indian Ocean and the Future of American Power*, New York: Random House, 2011.

plies, and this deficiency needs to be remedied.[22] In other words, energy security is intimately tied to Chinese strategic thinking and military posture, and therefore to the future strategic and military environment in East Asia.

When one examines these propositions in greater depth, it becomes clear that energy security is just one of several factors determining Chinese strategic and military decisions—and alleviating China's energy insecurity or changing its energy mix (between oil, gas, coal, and nuclear) will not necessarily lead to reduced instability and military competition in East Asia.

China's Military Modernization Program

Although "energy security"—and the threat of sea-based interdiction of supply—has been an official concern since 2003, there is little evidence to suggest that such concerns are primarily driving Chinese force posture and modernization over the past decade.

For a start, the 2006, 2008, and 2010 *Defense White Papers* only briefly alluded to energy and oil security as one amongst many factors. In the chapter on "Supporting Economic and Social Development" in the 2013 version, it merely states that the PLA exists, inter alia, to offer "security support for oil and gas exploration," and merely notes that "Security issues are increasingly prominent, involving overseas energy and resources, strategic sea lines of communication (SLOCs)…."[23] The point is that PLA Navy (PLAN) assets will be used to provide greater security now and into the future for Chinese energy interests. But energy security concerns are not primarily driving force posture as such. Indeed, it is clear that the

22. See, Saira Basit and Oystein Tunsjo, "Emerging naval powers in Asia: China's and India's quest for sea power," *Oslo Files on Defense and Security,* no. 2/2012, Oslo: Norwegian Institute for Defense Studies, June 2012.

23. *The Diversified Employment of Chinese Armed Forces,* Information Office of the State Council, People's Republic of China, April 2013.

shift beyond "near coast defense" in developing a blue-water navy is primarily about "safeguarding national sovereignty, security and territorial integrity."[24] That Chinese claims in the East and South China Seas are not primarily about claiming seabed energy resources will be discussed shortly.

There is no doubt that these new "historic missions" as former President Hu calls them encompasses considerations about energy security and China's access to SLOCs. But Chinese defense planners realize that there is simply no way to adequately solve its Malacca Dilemma. For example, the Malacca Straits is seemingly beyond the PLA's emerging capabilities designed to deny the U.S. sea-control. At best, the PLA will be able to only defend coastal SLOCs for the next several decades through inflicting "prohibitive costs" on U.S. ships operating in these coastal areas. Even if the PLA eventually "neutralizes" the Malacca Dilemma, China is powerless to stop U.S. blockading other potential energy-related chokepoints such as the Gulf of Aden or Straits of Hormuz—and will lack the capacity to do so for many decades beyond 2030.

Indeed, whether a U.S. blockade of oil supplies into China is a realistic option at all, and therefore a plausible concern for China, will be discussed below.

Is Blockade a Plausible and Genuine Concern for China?

Although writings by some Chinese strategists are replete with concerns about U.S. blockade,[25] more sensible and measured strategists and officials recognize that such a scenario is highly unlikely if not inconceivable. The first thing to note is that such an American action

---

24. Ibid.

25. See, for example, a summary by Andrew Erickson and Lyle Goldstein, "Gunboats for China's New 'Grand Canals'? Probing the Intersection of Beijing's Naval and Oil Security Policies," *Naval War College Review* 62, no. 2, Spring 2009, pp. 43-76.

would be illegal according to international and trade law, and would be only contemplated in the event of a major war against China. In other words, any blockade is an extreme action and is itself an act of war against China. The United States would suffer significant diplomatic fallout were it to attempt the blockage of China outside the confines of a major war. Indeed, U.S. attempts at a blockade would not be supported by its allies unless a full-scale war was occurring, due to the economic fallout that would invariably occur from disruption to the Chinese economy and turmoil in petroleum markets affecting the global economy. Put differently, the threat of blockade is only conceivable after a violent breakdown in relations between the U.S.-China; it is not a peacetime threat.

Second, it is highly dubious that a blockade could actually succeed. If the Malacca Straits were closed to China, there would be other deep-water passages through the Indonesian archipelago available to China.[26] Even if the United States managed to block these routes (e.g., through the Lonbok, Makassar, and Mindoro Straits) tankers could re-route around Australia—causing an acceptable 4-16 day delay for Chinese ports according to expert estimates.[27] Furthermore, even if the United States decided to impose a more "upstream" blockade in the Gulf of Aden or the Straits of Hormuz, this would not prevent energy shipments reaching China from Africa or South America. Because oil is a genuine global commodity, China would be able to obtain supply from non-blockaded sources.

Third, there would be an operational problem. As a globally traded commodity, it is very difficult to know where oil from Africa or the Middle East actually ends up. As a tanker leaves the Middle East towards Asia, its final destination can change a dozen times while

---

26. See Michael McDevitt, "The strategic and operational context driving PLA Navy building," in Roy Kamphausen and Andrew Scobell, eds., *Right Sizing the People's Liberation Army: Exploring the Contours of China's Military*, Carlisle, PA: Strategic Studies Institute, 2007.

27. See Gabriel B. Collins and William S. Murray, "No Oil for the Lamps of China?" *Naval War College Review* 61, no. 2, Spring 2008, pp. 35-58.

it is at sea, based on ongoing trading for the commodity. Tankers also often carry oil for customers in multiple countries, while bills of landing can be forged or fabricated to bypass blockages.[28] It would be an almost impossible task for one country (and its allies) to monitor and inspect all ships passing through chokepoints.

Furthermore, even if the United States somehow successfully blockaded Chinese bound oil tankers, Chinese NOCs could still purchase oil contracts on commodity markets and tankers already in the South China Sea could re-route to Chinese ports. In other words, unless the United States imposed direct control over oil commodity markets and imposed a market ban against supply to China—a clear impossibility—a blockade is unlikely to succeed. Besides, as the 1973 Arab oil embargo demonstrated, the embargoed country can simply purchase oil from third parties, even if it is at increased prices.

Fourth, a blockade assumes acquiescence and cooperation from key countries: Singapore, Malaysia, and Indonesia into the South China Sea; Saudi Arabia in the Middle East. For the Southeast Asian countries, free and unimpeded trade is their lifeblood. For Saudi Arabia, selling oil on commodity markets is critical to national income and therefore regime security. In 2013, the Saudi government received around $375 billion in revenue from oil exports, up from about $179 billion in 2010.[29] Oil imports from Saudi Arabia to China have more than doubled over the past five years and currently stand at over one million bpd.[30] It is difficult to envisage a scenario extreme enough for these countries to support an American blockade against China.

---

28. Ibid.

29. See "Saudi Arabia's Dependency on high oil prices courts disaster," *South China Morning Post*, November 13, 2013, available from *www.scmp.com/business/commodities/article/1354720/saudi-arabias-reliance-high-oil-prices-courts-disaster*.

30. See Michael Lelyveld, "China Ups Persian Gulf Imports," *Radio Free Asia*, February 4, 2013, available from *www.rfa.org/english/commentaries/energy_watch/oil-02042013105305.html*.

Are East Asian (Especially Chinese) Maritime Claims Primarily about Energy Security?

Strategists from East Asia and the United States have frequently asserted that interest in sea-bed resources is a major driver of growing interest in disputed regions of the East and South China Seas. However, talk about potential "resource conflicts" obscures the fact that the increasingly strident Chinese claims in these areas are more driven by considerations of political legitimacy and nationalism than it is about energy security. This is so for several reasons.

For a start, much confusion is caused by the failure to distinguish between hydro-carbons in "uncontested coastal areas" (which are already being mined for hydro-carbons), "contested coastal areas" (where disputes over bidding blocks are the primary issue of contention), and more distant and contested deep-water areas with largely unknown hydrocarbon potential.

Reports drawing from Chinese speculation that there could be 200 billion barrels of oil in the South China Sea—making the region a mini-Saudi Arabia—are highly misleading for a number of reasons.

First, this is a completely unverified estimate and there is no evidence that officials in Beijing actually believe the estimate to be even vaguely accurate (even though the estimate has been repeated by official bodies).[31] The 200 billion barrel figure is produced by a fundamentally unsound extrapolation of what may lie in the contested areas of the South China Sea that simply takes the known reserves in uncontested coastal zones (about 11.2 billion barrels of oil and 190 trillion cubic feet (Tcf) of gas) and assumes that contested areas will be as similarly rich.

Besides lacking credibility, the 200 billion barrel figure is based on what is beneath the sea-bed, not what is actually recoverable. In compiling a "best guess" of recoverable resources in the contested

---

31. For example, see, Will Rogers, "Beijing's South China Sea Gamble," *The Diplomat*, February 4, 2012, available from *http://thediplomat.com/2012/02/beijings-south-china-sea-gamble/*.

zones, experts offer the figure of 1.6-6 billion barrels, which would deliver about 650,000 bpd for a decade before declining sharply when based on standard deep water depletion curves.[32] Given that China currently imports over 4 million bpd, which will rise to around 10 million bpd by 2020, even the best case estimate for contested areas in the South China Sea will not provide oil security for China.

Second, the deep water, strong undersea currents, and prevalence of tropical storms in these areas will make extraction enormously difficult and expensive, if not impossible. In any event, given that affordability is an essential element of Chinese energy security, the expense of extracting deep water oil would likely render it prohibitive. Indeed, mining of seabeds is much more feasible in the shallower waters of uncontested areas of the South China Sea. When one looks at the East China Sea, the importance of energy in the dispute between Japan and China is even less pronounced. While some Chinese sources predict some 160 billion barrels of oil and 250 Tcf of gas, International Energy Agency estimates are closer to 60-100 million barrels of oil and 1-2 Tcf of gas.[33] With a record of consistently over-stating energy repositories in contested regions such as Xinjiang and the South China Sea for domestic political reasons, there is little evidence that sensible Chinese officials actually rely on their own elevated estimates. Once again, there is no estimate of recoverable oil and gas amounts. In any event, there is little prospect that China and/or Japan would initiate a highly costly and disruptive resource conflict (which would likely involve American forces) over a region whose seabed resources are almost entirely unproven and speculative.

In summary, even if realistic best case scenarios of recoverable re-

---

32. See Nick Owen, "Oil Disputes in the South China Sea in Context," in Clive Schofield, ed., *Maritime Energy Resources in Asia: Energy and Politics*, NBR Special Report, no. 35, Washington DC: National Bureau for Asian Research, December 2011, pp. 31-35.

33. See U.S. Energy Information Administration, "East China Sea Analysis Brief," September 4, 2012.

sources for the East and South China Seas are assumed, a successful Chinese attempt at colonizing these entire domains and extracting its resources would deliver China only a few years' worth of oil and a decade or more of gas needs.[34]

Instead, resource issues and appeals to "national energy security" are generally used to harden domestic resolve in claiming these territories. If extracting resources were the primary motivation, China would be far more prepared to engage in joint ventures with other claimants in order to provide a political opening through which it can exploit its share of the resources—something China has generally resisted.

These worsening disputes in the East and South China Seas are therefore better understood as a function of rising nationalism. They are also a function of structural conflict (caused by China's rise outside the Western alliance system) and geostrategic competition within which China feels acutely uncomfortable with the American and allied military presence within in its First Island Chain.[35]

Relying on the American Seventh Fleet to preserve the stability and openness of SLOCs is part of China's broader strategic dilemma: It continues to free ride under the American security umbrella because it has no choice but to do so since no other country is capable of fulfilling the role of ensuring stability and providing order for commerce to take place. At the same time, structural and strategic competition is deepening between China on the one hand, and the United States and its allies on the other.

34. See, U.S.-China Economic and Security Review Commission, "China's Maritime Disputes in the East and South China Seas," Opening Statement of Dr. Steven W. Lewis, Hearing before the U.S.-China Economic and Security Review Commission, , 113th Cong., 1st sess., April 4, 2013, Washington: DC, U.S. Government Printing Office, 2013, p. 108.

35. For example, see, Aaron Friedberg, *A Contest for Supremacy: China, America and the Struggle for Mastery in Asia*, New York, W.W. Norton & Co, 2012, for an argument on the inevitability of structural tension between the two countries.

Implications for East Asian Energy Futures and Markets

Japanese force posture since the end of the Second World War is limited by a coastal defense doctrine while South Korean force posture is largely determined by tensions with North Korea in the Peninsula. Even though China has a more expansive military and naval force posture and doctrine, this paper has so far argued that one should not overstate the role of energy security in shaping this posture and doctrine. A stated desire to take on more responsibility in patrolling the SLOCs that its energy imports depend upon does not equate to strategic planning being defined primarily by energy security concerns. More broadly, one should not over-emphasize the role of energy in shaping the strategic and military future of East Asia.

One critical implication of all this is that energy self-sufficiency—even if it were possible for East Asian countries—is not likely to dampen geostrategic and military competition between them. The preceding section is also provided as a counter to the common assumption that "resource wars" are primarily driven by scarcity when such wars are caused by political and economic policies (in response to scarce resources) rather than scarcity itself.

None of this is to imply that the future shape of energy policies and energy markets is irrelevant, or that we should remain complacent.

On the contrary, in order to ensure that energy policies do not contribute to a worsening of existing structural and geostrategic tension, one ought to keep three things in mind:

1. As argued earlier, all major East Asian states, especially China, have adopted a securitized approach to energy security, in that supply of energy is seen as a core element of national and regime interest. However, none of the major East Asian states have abandoned a market-first view in favor of a militarized-first view of energy security—for the reason that the latter would actually lead to far more inefficient energy supply and price outcomes. For example, China's offshore equity assets would only provide it with

one-quarter of its import needs if these offshore sites were utilized at full capacity. Even now, less than 20% of offshore oil owned by its NOCs is shipped back to China.

If any of the major East Asian states were to take a military-first view of energy security—with the effect of increasing the energy insecurity of one's own country and that of the region—then existing geostrategic tension and competition could worsen and possibly spiral towards a violent scenario. Being outside the American alliance system and an entrenched political-economy that is intimately tied to the CCP, this paper argues that China is the most likely East Asian power to revert to a "self-help" energy security posture. However, it has strong pragmatic and economic reasons to resist that option, which would lead to worse energy security outcomes for itself and the region.

2. Decreasing a militarized view of energy means increasing (not decreasing) reliance on regional and global commodity markets, and also seaborne trade which serves as a restraint on the prospect of naval conflict. A loss of faith in commodity markets would invariably lead to a much more militarized "self-help" system as states and governments increasingly adopt a zero-sum approach to energy security.

3. Paradoxically, in an era in which no East Asian state will gain sea control of the maritime commons (even if their sea denial capabilities are enhanced), increased reliance on seaborne energy imports will increase the vulnerability of energy importers—thereby making the costs of partial disruption of supply and volatility in prices which would result in the event of major war in the East or South China Seas increasingly prohibitive.

For Japan and South Korea, which import virtually all of its fossil fuels, and China, which imports more than half of its oil needs, the above logic is particularly compelling even if military planners in all three countries prepare for a more violent future.

What does all this mean for a set of desirable trajectories as to how future energy mixes and markets develop? The below are a set of suggested broad guidelines and suggestions.

A. A more commoditized and openly traded energy resource (from supply, to delivery, to market pricing) tends to lower the effectiveness of militarized approaches to energy security.

- It is far more difficult to "lock in" supply for oneself in commoditized and open markets. Indeed, as long as major suppliers remain committed to freely traded commoditized markets, clients in East Asia have no option but to rely on and trust these markets to meet energy demand.
- Blockades and other military actions are far less effective when supply and delivery of resources are determined by the operation of well-functioning markets.
- When prices are set by the market rather than being based on bilateral agreements, the leverage of dominant suppliers and/or dominant buyers is diluted—which will help to take the politics out of the energy trade.

B. In terms of energy mix for East Asian powers, the greater the future reliance on openly traded resources in commodity markets, the less militarized energy policies are likely to be.

- Oil markets are deeply commoditized. Natural gas markets, by contrast, are more fragmented than oil markets and countries like China seek greater vertical ownership and control (e.g., through off-shore equity or else exclusive supply deals, along the pipeline or a liquid natural gas transport chain). Countries are likely to seek greater control when it comes to pipeline gas as this is easy to disrupt, while seaborne trade is much harder to disrupt. This is where a lack of a commonly patrolled and supervised East Asian gas pipeline network is a problem.

- Nuclear currently provides about 1% of China's energy needs. Even if reliance on nuclear is increased significantly, China will still require an increasing amount of oil and gas imports from the Middle East, Africa, Southeast Asia, and Australia. In other words, China will remain dependent on oil and gas commodity markets regardless of the nuclear energy equation. Nuclear provides about 10% and 30% of Japan's and South Korea's energy needs respectively.

C. The greater the prevalence of private sector actors in markets (encompassing suppliers, intermediaries such as transporters, and buyers), the less confluence between geostrategic and political interest on the one hand, and energy policy on the other—therefore, the less securitized energy policy becomes.

- The more interventionist the state is in the economy, the greater the role of the government in energy policy tends to be since energy security is considered too important to be left entirely alone to "market forces."[36] A nationalistic approach seeks to use state-owned-enterprises (SOEs) to own and control energy across the vertical spectrum from exploration and extraction, refinement, transportation, and consumption—with military support and protection offered to the interests and activities of these SOEs.

- Energy SOEs tend to be far less efficient than Western

36. See John Lee, "China's Geo-Strategic Search for Oil," *Washington Quarterly* 35, no. 3, Summer 2012, pp. 75-92.

multinational energy companies meaning that countries like China actually rely on efficient private sector firms to meet demand from global markets. Even if East Asian SOEs resist privatization (and the momentum for this largely depends on continued economic reform in countries such as China,) the genuine commercialization of energy SOEs would be a step in the right direction. This decreases the prospect that they will be used to pursue national strategic and political interests and strengthens the development of better functioning commodity and market-driven energy sectors. In this context, companies such as Malaysia's state-owned company, Petronas, serves as a good model for Chinese SOEs.

D. The greater the volume and diversity of players committed to market-based supply of fossil fuels, the better.

- This is where America has a potentially critical role to play into the future. If America becomes a major net exporter of energy (gas and oil) over the next two decades, it is critical that private sector energy companies are allowed to sell to export markets based on commercial considerations alone, and eschew strategic or political reasoning that may be favored by Washington.

  This will have one of two beneficial effects. First, it will provide extra buffeting against political shocks and risks that might afflict some energy producing areas and nations, hence alleviating the energy insecurity and vulnerability of East Asian countries who might otherwise take a much more self-help view of meeting energy needs in order to insure against such shocks.

For example, the increased American oil production since 2008 of around 2.7 million bpd amounts to three times the total oil production lost to world markets as a result of the Arab Spring.[37]

Second, it will reduce the capacity of oil and gas exporters such as Russia to use energy exports as a political weapon. It will also lower East Asian reliance on "rogue states" such as Iran.

E. **The more buy-in from "outsiders" such as China into regional and global organizations such as the International Energy Agency, the more institutionalized open energy markets become—leading to reduced securitization.**

- This is a "chicken versus egg" problem: China has refused invitations to join the IEA as an associate member as it believes that releasing trade and energy data will jeopardize its national strategic and energy interests.

Conclusion

China will remain wary of the American and allied naval presence in East and Southeast Asia due to inevitable geostrategic tensions triggered by China's rise and deep discomfort with American strategic staying power in the region. While not decisive, energy policies and mindsets can either exacerbate these tensions or serve as a restraint from such tensions escalating into conflict.

In particular, the deepening of collective reliance on an open market based approach to securing energy imports will mean that China will have little option but rely on the American naval presence to supervise the SLOCs used to bring energy imports into the region since no other country will have the capability to fulfill this role in the foreseeable future. In this context, Beijing is more likely to continue to grudgingly accept that the American presence is essential

---

37. See, Amy Myers Jaffe and Edward L. Morse, "Liquefied Natural Profits," *Foreign Affairs*, September 16, 2013, available from *www.foreignaffairs.com/articles/united-states/2013-09-16/liquefied-natural-profits*.

as a stabilizing force in the region needed to ensure uninterrupted supply of energy resources to China—even if the PLA Navy takes on more responsibility for patrolling SLOCs in the Indo-Pacific.

Washington has a critical role to play here in de-securitizing the operation of energy markets and how energy resources are bought and transported to China, particularly if it emerges as a significant energy exporter to the region. Remember that as far as China is concerned, its primary strategic competitor is also the self-assigned champion of open market systems in the region and world. While strategic competition between the United States and China will likely deepen over time, it is in America's and the region's interest to do whatever it can to convince Beijing that Washington is at least committed to supporting and protecting open market systems, universal access to SLOCs, and other essential economic processes that benefit all East Asian powers.

Moreover, creating incentives for the region to view energy security in economic rather than military terms will offer further encouragement for American allies such as Japan to see the security relationship as one required to preserve stability, freedom of commerce and rule of law required for energy and other trade, and less as a security arrangement that can be used to boost their escalatory options in bolstering their claims in disputed territories—something the United States has been trying to avoid whilst at the same time keeping its alliances robust. Indeed, greater collective reliance on SLOCs in the East and South China Seas is likely to strengthen regional convention and norms that "freedom of passage" for all commercial shipping is applied to SLOCs, even if the SLOCs pass through disputed maritime regions.[38]

None of this is any guarantee that tensions will subside between China and other powers in Asia. But, it means that the dire predictions about "resources wars" in the region are less likely to come to pass.

---

38. The issue of whether freedom of "navigation" which extends to military vessels is a more fraught question.

CHAPTER 2

Energy Futures, Market Signals, and Cooperation in Northeast Asia

Kyung-Jin Boo and Ji-Chul Ryu

This chapter, first, examines the current situation of energy supply and demand in Northeast Asia (NEA). Next it continues to introduce the long-term projection of energy supply and demand in the region based on an assumption of major economic and energy indicators such as gross domestic product (GDP), population, crude oil price, etc. After overviewing the energy situation and the long-term energy projection up until 2035 followed by identification of issues related to energy supply and demand in the NEA region, it moves to a detailed and sincere discourse on energy cooperation in the NEA region.

The Future of Energy in Northeast Asia[1]

Republic of Korea

Energy in General.[2] The Republic of Korea (ROK or South Korea) has no domestic oil resources and has produced only a small amount of anthracite coal, forcing it to import most of its coal, which is bituminous coal. Consequently, South Korea has to import 97% of its energy needs and is ranked as the fifth-largest oil importer and the second-largest importer of liquefied natural gas (LNG) in the world. Although total primary energy consumption is dominated by oil and coal, nuclear power and LNG also supply a significant share of the country's primary energy consumption.

Total primary energy consumption was 278.7 million [metric] tons of oil equivalent (MTOE) in 2012, increasing by 5.1% a year since 1990. The most growth occurred in natural gas (13.6%) and nuclear (4.1%). Oil use increased at a relatively slower rate of 3.5% a year during the same period. Total final energy consumption (TFEC) in 2012 was 208.1 MTOE, increasing at an average annual growth rate of 4.7% from 1990. The industrial sector accounted for 61.7% of final energy consumption in 2012, followed by residential/commercial (18.2%) and transportation sectors (17.8%). Consumption of natural gas in the industrial sector shows the fastest annual growth rate of 18.7% during the same period and oil accounts for a rela-

---

1. This section is solely quoted from the Economic Research Institute for ASEAN and East Asia's (ERIA) recent work in which the author, Boo was involved as a member of the Working Group on Energy Saving Potential of ERIA. See the group's report, Shigeru Kiruma, ed., "Analysis on Energy Saving Potential in East Asia," ERIA Research Project Report 2012, no. 19, Jakarta, Indonesia, June 2013, available from *www.eria.org/publications/research_project_reports/FY2012-No.19.html*.

2. Numbers and figures are based on energy statistics in Korea Energy Economics Institute (KEEI), "2013 Energy Info. Korea," Ulsan, Republic of Korea: KEEI, December 2013, available from *www.keei.re.kr/web_keei/en_publish.nsf/by_energy_info/710D6CBD0A2326D149257CCC004B1EE3/$file/EnergyInfo2013.pdf*.

tively large share of industry consumption.

In 2012, electric power generation in South Korea reached 530.6 terawatt-hours (TWh) of electricity, with coal and nuclear combined providing way over two-thirds of South Korea's electricity. Natural gas accounted for 22.4% of electricity generation in 2012. Total electricity consumption grew at an average annual rate of 7.3% over the period from 1990 to 2012. When broken down by fuel type, coal, natural gas, and nuclear have grown by an average annual rate of 11.0%, 11.9%, and 4.7% respectively over the same duration. Such a rapid growth rate in electricity generation was due to the high growth in the GDP per capita electricity consumption, quadrupled from 2,202 kilowatt-hour per person (kWh/person) in 1990 to 9,331 kWh/person in 2012.

Pursuant to the Basic Energy Law, the South Korean government has established and implemented National Energy Basic Plans that are supposed to be conducted every five years over a period of 20 years. The purpose of each plan is to suggest the direction of future-oriented energy policies and determine mid and long-term strategies to systematically secure energy resources, expand stable infrastructure for supplying domestic energy, and rationalize the use of energy needed for the sound development of the national economy. The plan should also include policies to minimize energy-related factors that harm the environment and also to effectively contribute to the achievement of national energy policies for expediting the development of energy related technologies.

Long-Term Energy Outlook. South Korea's final energy consumption showed a growth of 4.4% per annum from 64.9MTOE in 1990 to 157.4 MTOE in 2010.[3] The non-energy sector had the highest growth rate during this period at 9.1% per annum followed by the industry sector with 4.3% growth. Energy consumption in the residential/commercial/public (other) sectors grew at a relative-

---

3. The energy consumption figures are based on the net calorific values as converted by the Institute of Energy Economics, Japan (IEEJ) from original data submitted by the Republic of Korea.

ly slow pace of 3.1% per annum.

With an assumption of low economic and population growth, final energy consumption in the ROK is projected to increase at a low average rate of 1.5% a year between 2010 and 2035. This stems largely from the slow growth in energy consumption in the transportation sector. The strongest growth in consumption is projected for the industrial sector with an average annual rate of 1.9% between 2010 and 2035.

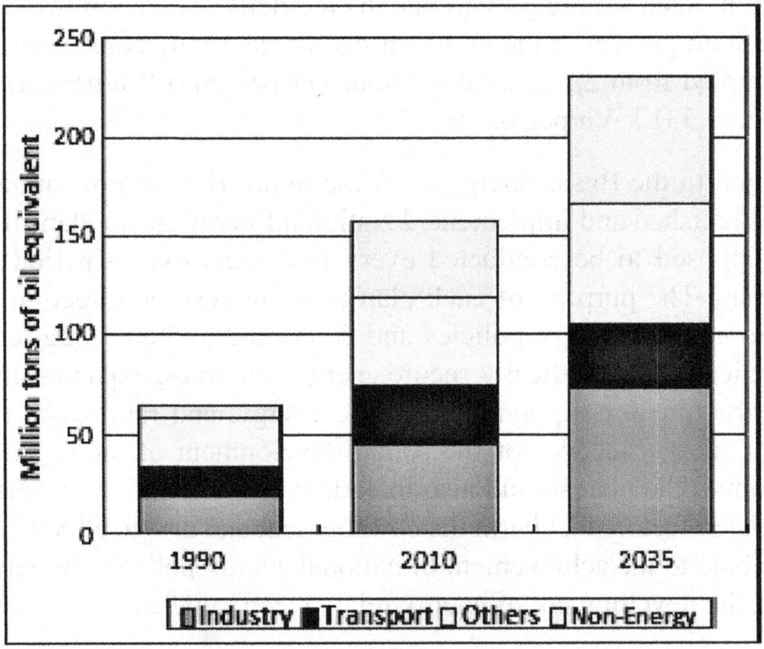

Figure 1A. Final Energy Consumption by Sector.

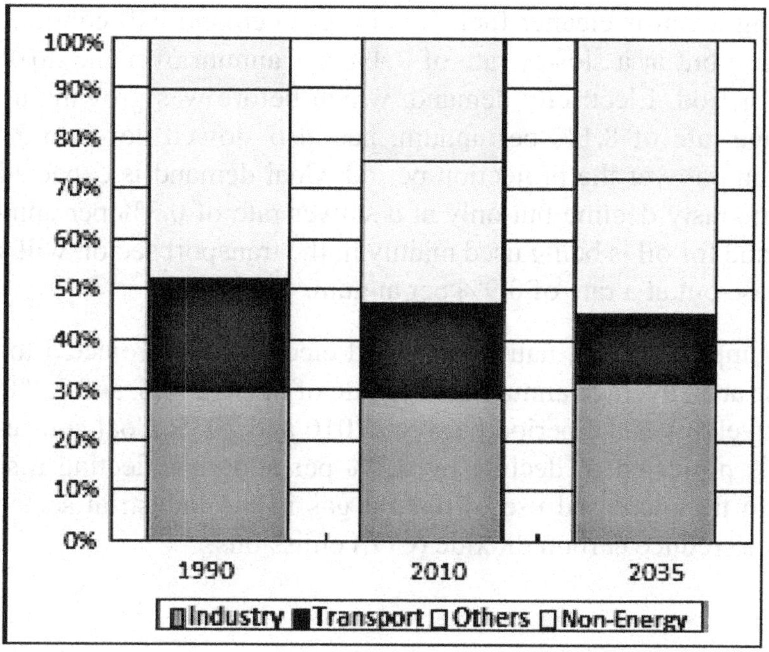

Figure 1B. Final Energy Consumption by Sector.

In 1990, the residential/commercial/public (other) sector had the largest share in the total final energy mix of South Korea (37.5%). In 2010, the share of this sector declined to around 28% due to the rapid increase of demand in the non-energy sector. The industry and transport sectors also experienced a declining share in the total final energy mix over the same period. The industry and non-energy sectors will grow faster than the transport and other sectors. Consequently, the share of these sectors will increase over the projection period. The industry sector share will increase from 28.4% in 2010 to almost 32% in 2035. The non-energy sector share will increase from 24.4% to 28% over the same period.

The demand for natural gas in the end-use sector has increased significantly over the 1990 to 2010 period, increasing at an average annual growth rate of 18.6%. Coal, on the other hand, declined at an average rate of 1% per annum reflecting that South Korea is moving towards cleaner fuel. Natural gas demand will continue to increase but at a slower rate of 2.4% per annum over the 2010 to 2035 period. Electricity demand, which before was growing at an average rate of 8.1% per annum, has also slowed down to 2.4% per annum over the projection period. Coal demand is expected to continuously decline but only at a slower rate of 0.3% per annum. Demand for oil is being used mainly in the transport sector, will still increase but at a rate of 0.9% per annum.

Consumption of oil, natural gas, and electricity is projected to increase at an average annual growth rate of 0.9%, 2.4%, and 2.4% respectively over the period between 2010 and 2035. Coal consumption is projected to decline by 0.3% per annum reflecting a shift toward the increased use of natural gas in the industrial sector in order to reduce carbon dioxide (CO_2) emissions.

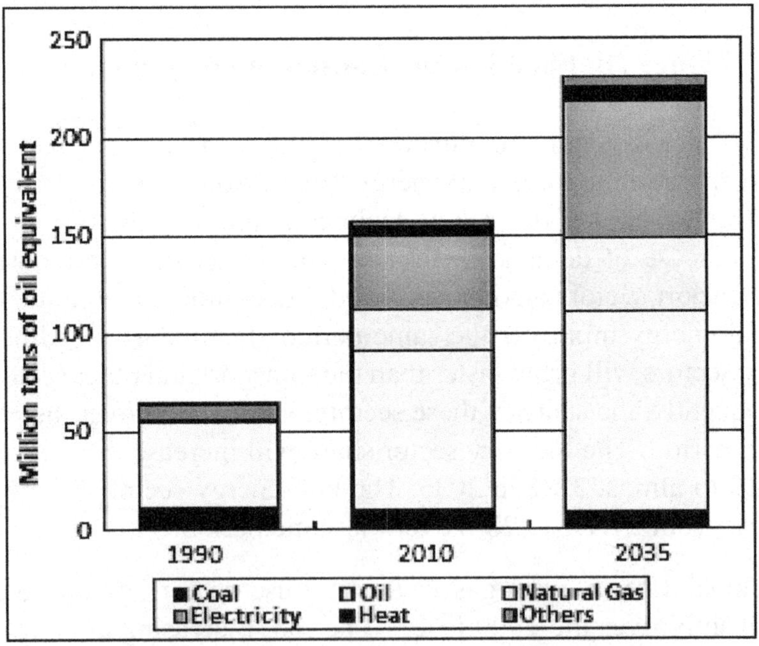

Figure 2A. Final Energy Consumption by Fuel.

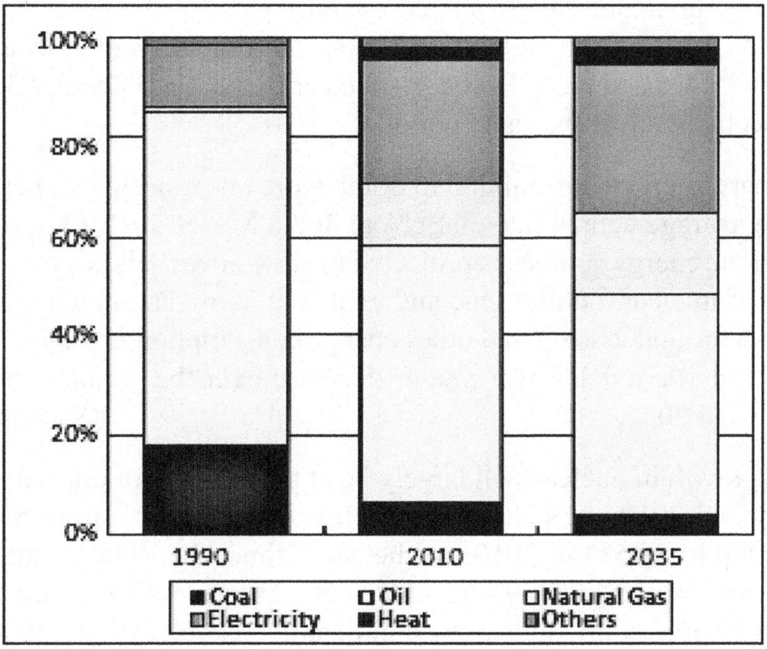

Figure 2B. Final Energy Consumption by Fuel.

Oil was the most consumed product with a share of 67.3% in 1990 but declined to 52.0% in 2010. The share of coal in the final energy consumption has declined by 12% a year between 1990 and 2010 whereas the share of electricity has doubled to be the second largest consumed product. Natural gas, with its rapid growth, has an increasing share from 1% in 1990 to almost 13% in 2010.

In the future, the share of oil will continue to decline. It is expected that oil's share will decline to 44.6% by 2035. Coal consumption will also decline and be replaced by natural gas. The share of coal will reach 3.8% in 2035 as compared to 6.1% in 1990. Natural gas, on the other hand, will increase its share from 12.9% in 1990 to 16.2% in 2035. Electricity share in the total final energy will also increase from 20.4% to 30.3%.

Primary energy consumption in South Korea grew at an average rate of 5.0% from 92.4 MTOE in 1990 to 247.3 MTOE in 2010.

Among the major energy sources, natural gas was the fastest growing at an average annual rate of 14.2%. In contrast, coal grew at the rate of 5.4% a year, followed by nuclear and oil at 5.3% and 3.3%, respectively, over the same period.

Primary energy consumption in South Korea is projected to increase at an average annual rate of 1.6% to 367.8 MTOE in 2035. Growth in all the energy sources is projected to slow down. While consumption of nuclear, natural gas, and coal will show an annual growth rate of around 2%, oil and other energy consumption is projected to increase at a much lower rate or decrease over the period between 2010 and 2035.

The growth in nuclear will largely be at the expense of oil. Between 1990 and 2010, oil's share had declined almost 30%, from 53.9% in 1990 to 38.5% in 2010. At the same time, the share of nuclear had increased from 14.9% to 15.7%. Natural gas's share in the total primary energy mix increased significantly between 1990 and 2010, from 2.9% to 15.6%. Coal, however, increased slightly from 27.7% to 28.7%. The share of oil will continue to decline to around 32% by 2035, while nuclear's share will increase to 17.5%. Natural gas' share in the total mix will be similar to that of nuclear while coal's share will be almost equal to that of oil. Hydro's share in the total energy mix will remain constant.

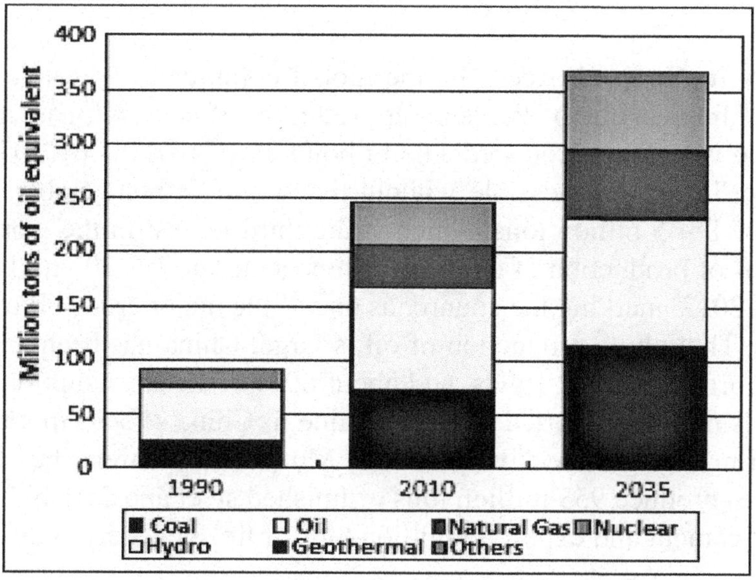

Figure 3A. Primary Energy Consumption.

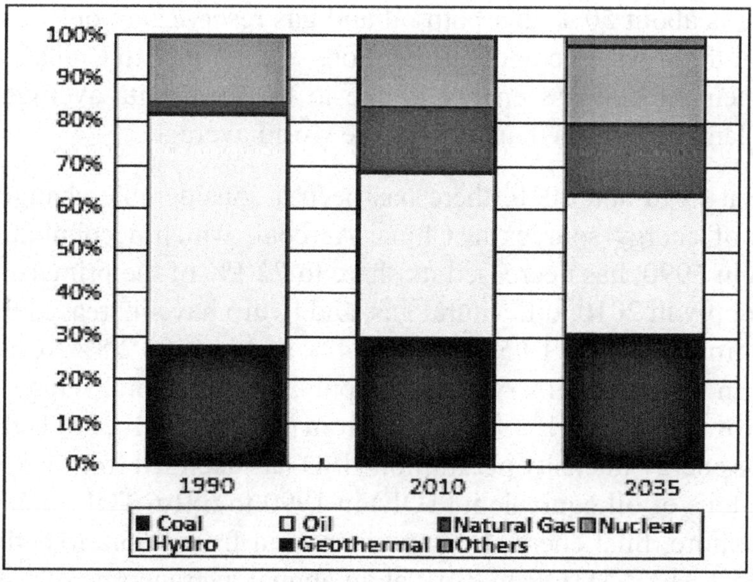

Figure 3B. Primary Energy Consumption.

China

Energy in General. Recently, the global community was able to notice China as one of the leading producers of coal, oil, gas, and hydropower. China produced 3.65 billion tons of coal in 2012. This may not have been possible without the country's recoverable reserve of 114.5 billion tons, which is the third largest in the world. In terms of production of crude oil, China's output was 207 million tons in 2012, marking the country as one of the major crude oil producers. Though its production of oil is large, China has been a net oil importer since the 1990s, and about 60% of oil consumption in China is met by imported oil. This is due to China's focus in manufacturing energy-intensive products. Moreover, the trend has led China to produce 953 million tons of finished steel and 2.21 billion tons of cement and export 56 million tons of finished steel in 2012.

When it comes to describing per capita resource endowment, China falls very short of the world average in per-capita energy reserves. China's per-capita energy reserve of both coal and hydropower resources is about 50%, and both oil and gas reserves are only about 6.6% of the world average. Furthermore, a hindrance to China's development of biomass energy is due to the per-capita average of arable land being less than 30% of the world average.

Between 1990 and 2010, there has been a considerable change in shares of energy sources in China. As coal, which accounted for 78.7% in 1990, has decreased its share to 72.1% of the primary energy supply in 2010, oil, natural gas, and hydro have increased their shares from 17.8% to 19.5%, 1.9% to 4%, and 1.6% to 2.8%, respectively. In terms of primary energy supply, it grew at an average annual growth rate of 6.1% since 1990 from 671.7 to 2,212.5 MTOE in 2010 as energy intensity per million USD has declined from 1,510.8 metric tons of oil equivalent (TOE) in 1990 to 681.6 TOE in 2010. Furthermore, final energy demand increased from 466.2 MTOE in 1990 to 1,312.7 MTOE in 2010 at an annual average rate of 5.3%. An increase in final energy demand didn't necessarily mean there was an increase in the share of final energy demand in coal. In 1990 coal accounted for 68.6%, but declined to 39% in 2010. This implies that other resources have increased in shares of final energy demand: Oil consumption rapidly increased its share at 7.6% per annum from 18.3% in 1990 to 28.1% in 2010. Furthermore, as industry, residential, and transport sectors became more prosperous, electricity demand increased at an annual rate of 10.7%. Its share in the final energy demand has increased faster than any other final energy sources—from 8.4% in 1990 to 22.6% in 2010.

In China, the industry sector consumes the largest chunk of energy, followed by the residential, commercial, and transport sectors. From 1990 to 2010, the industry sector's share of energy consumption had increased from 52.2% to 54.2%. However, due to rapid growth in the industry and transport sectors, energy consumption in residential and commercial sectors decreased to 21.9% in 2010 from 31% in 1990. Furthermore, coal-fired power generation accounted for 71.3% of total electricity generation in 1990 and by 2010 its share had increased to 77.8%. Other than coal and electricity generation, hydropower's share decreased to 17.2% in 2010 from 20%in 1990. The energy demand share of gas and oil accounted for 1.9% collectively, whereas nuclear power increased to 1.8% in 2010.

Energy Projection up to 2035. Considering the assumption of lower economic and population growth in China, the growth of final energy demand in the coming 20 years is projected to slow down. The projection for the average annual growth rate in final energy demand is 3.1%, increasing from 1,313 MTOE in 2010 to 2,829 MTOE in 2035. Energy demand in the transportation sector is expected to be fastest at an annual growth rate of 4.0%, followed by 3.8% in commercial and residential ('other') sectors together. Lastly, industry sector's energy demand is projected to grow at a rate of 2.5% per annum. Figure 4 below shows final energy demand and share of consumption by sectors in China for the years 1990, 2010, and the target year, 2035, respectively.

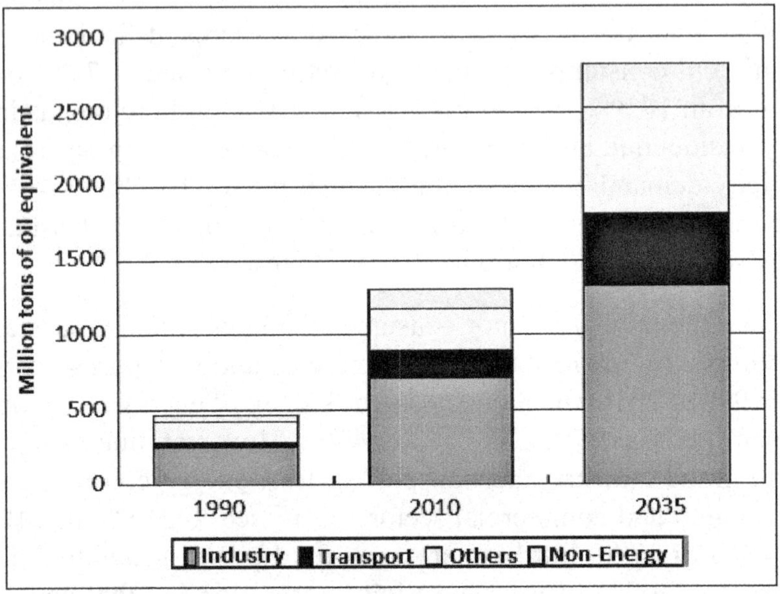

Figure 4A. Final Energy Demand by Sector.

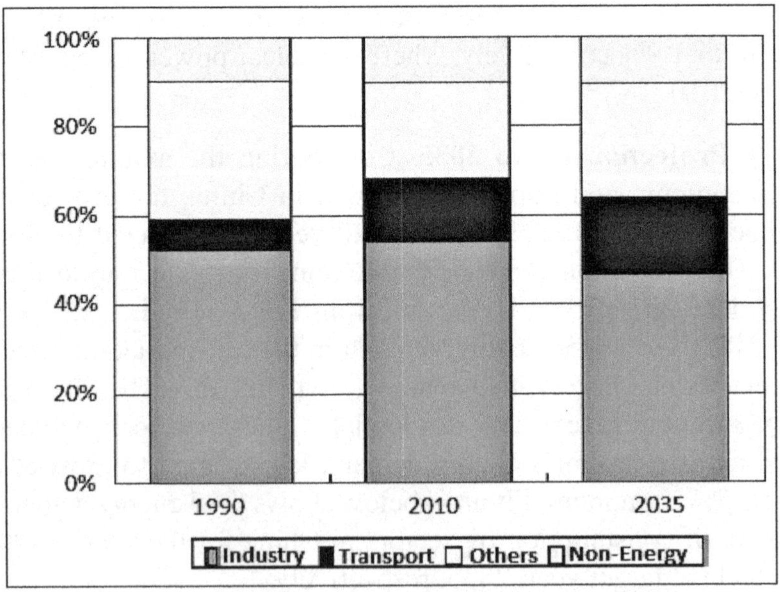

Figure 4B. Final Energy Demand by Sector.

Though coal has always held a large chunk of China's total final energy demand, much lower annual growth rate in its demand is expected in the next two decades, falling to 749 MTOE with an annual average growth rate of 1.5%, as compared with 2.4% between 1990 and 2010. Conversely, demand for natural gas is projected to show the most rapid growth of 7.9% per annum with an increase of 324 MTOE in two decades from 57 MTOE to 381 MTOE by 2035. Furthermore, demand for electricity, heat, and oil as final energy is projected to increase at an annual average growth rate of 3.7%, 2.5%, and 3.2% to reach around 732 MTOE, 120 MTOE, and 805 MTOE, respectively by 2035. Figure 5 below shows final energy demand and shares of energy sources in China for the years 1990, 2010, and the target year, 2035.

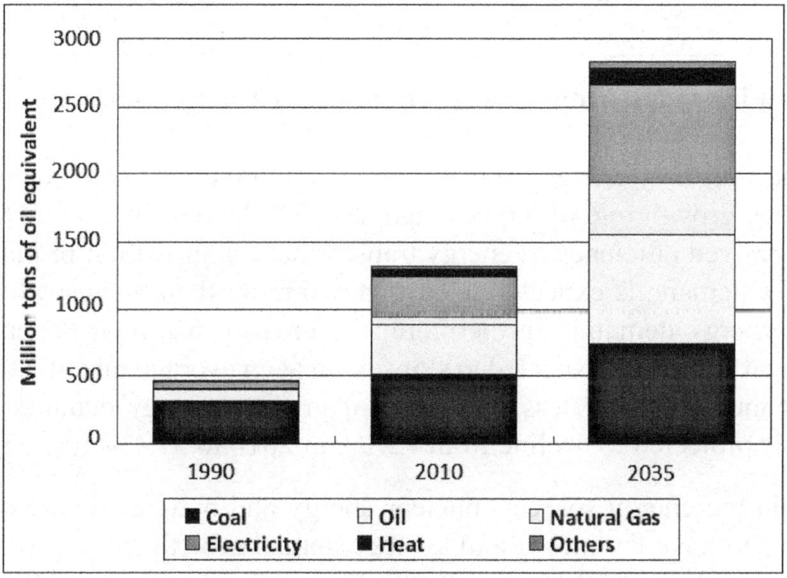

Figure 5A. Final Energy Demand by Energy Source.

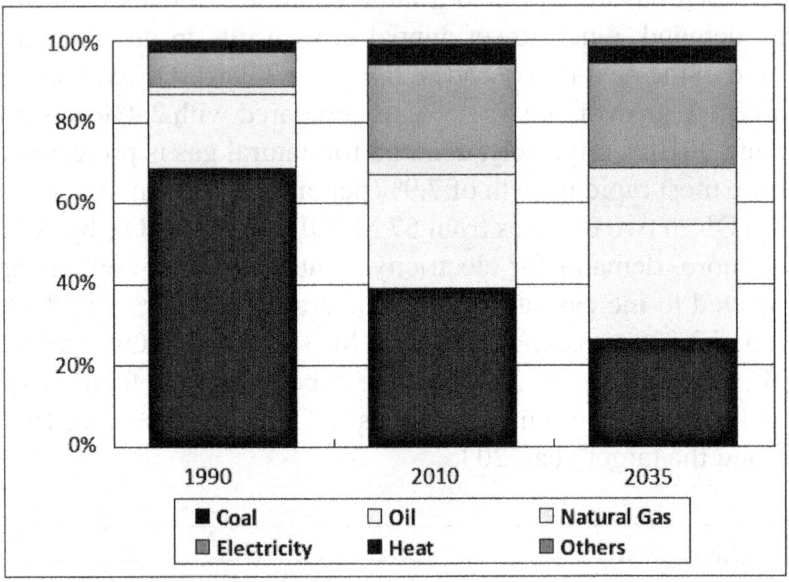

Figure 5B. Final Energy Demand by Energy Source.

China's primary energy supply is projected to increase at an annual average growth rate of 3.0%, reaching 4,585 MTOE by 2035. Due to improved efficiency in energy transformation, growth of primary energy demand is expected to slow down more than the growth of final energy demand. Furthermore, the growth in primary energy demand for coal is expected to slow down at an average rate of 2.0% per annum, nevertheless, the share of primary energy demand of coal is projected to decline from 72.1% in 2010 to 57.4% in 2035.

Within the energy sources, nuclear energy and natural gas are expected to have the most rapid average annual growth from 2010 to 2035 at a rate of 10.1% and 8.4% per annum, respectively. From 2010 to 2035, the ratio of natural gas and nuclear energy is projected to increase from 4% to 14.4% and 0.9% to 4.6%, respectively. Demand for oil and hydropower are projected to grow at an average annual rate of 2.7% and 3.3%, respectively, with a small decrease in the share of oil from 19.5% to 18.1% and a small increase in the share of hydropower from 2.8% to 3.0%. Figure 6 below shows primary energy supply and shares of energy sources in China for the years 1990, 2010, and the target year, 2035.

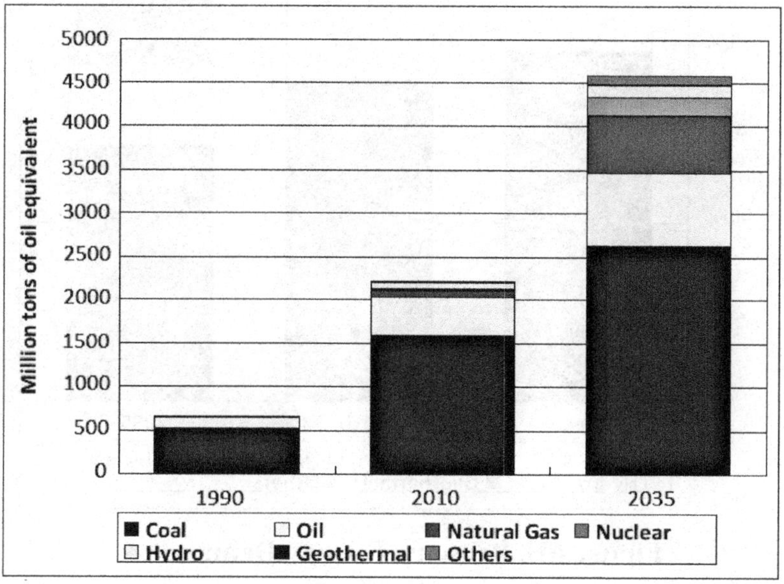

Figure 6A. Primary Energy Demand.

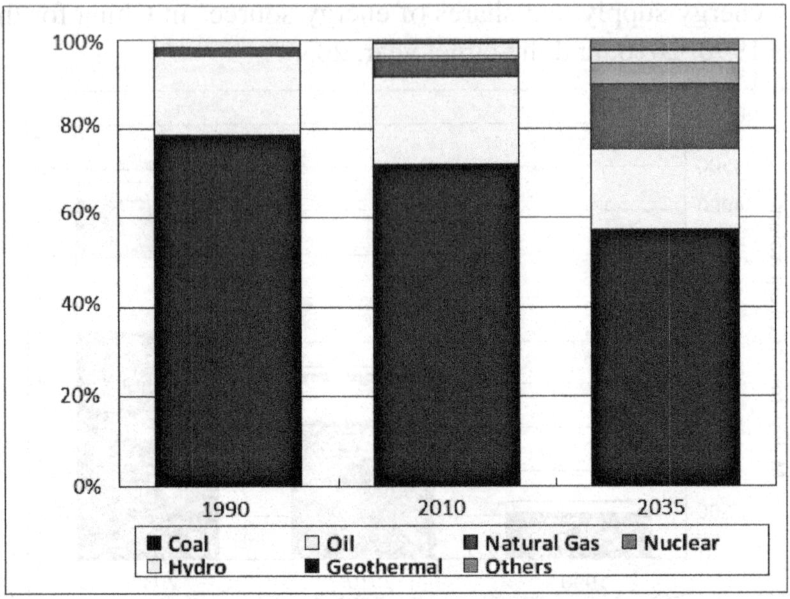

Figure 6B. Primary Energy Demand.

<u>Japan</u>

Energy in General. The primary energy supply in Japan was recorded as 494.0 MTOE in 2010 and consisted of 41.1% oil, 23.3% coal, 17.4% natural gas, 15.2% nuclear energy, with renewable sources such as wind, solar, geothermal, and hydro representing the remaining 3%. Furthermore, the net import of energy accounted for about 82% of the primary energy supply, where import dependence of oil, coal, and gas was 99%, 99%, and 96%, respectively.

Between 1990 and 2010, primary energy consumption in Japan grew at a rate faster than the rate of growth in final energy consumption at an average annual growth rate of 0.6%. Among those energy sources, the most rapidly growing energy sources were natural gas and coal. Natural gas, coal, and nuclear energy consumption grew at an average rate of 3.4%, 2.0%, and 1.6% per annum, whereas average annual growth for oil consumption slowed down to 1.0% from 1990 to 2010.

In 2010, 1,111 TWh of electric power was produced in Japan through an installed capacity of 282 GW. The fuel to generate electric power is broken down into following shares: 64% thermal, which consists of coal, natural gas, and oil, 28% nuclear, 7% hydro, and 3% renewable energy sources, such as solar, wind, and geothermal. Though nuclear itself has a very high share of electric power generation capacity, Japan had planned to increase this number by constructing and operating an additional 14 nuclear power plants by 2030. However, due to the Fukushima Nuclear Power Plant accident caused by the Great East Japan Earthquake in 2011, it is uncertain how many of the planned nuclear power plants will materialize by 2035. Hydropower is expected to increase in installed capacity, harnessing 70% of the resource potential by 2035. Power generation fueled by natural gas is also expected to increase more than any other fossil fuels. On the other hand, oil is projected to decrease in use by 2035.

Energy Projection up to 2035. Through the period between 2010 and 2035 Japan's final energy demand is projected to decrease at an average rate of –0.4%, considering the patterns of relatively low economic and population growth. The decline in final energy demand is further affected by less use of fossil fuels in transportation as well as by improving energy efficiency and use of renewable energy sources to power residential and commercial energy needs. However, between 2010 and 2035 the final energy demand in the industry sector is expected to grow at an average rate of 0.2% per annum. When looking at the fuel type consumption for the period between 2010 and 2035, natural gas and electricity are projected to increase at a slow average rate of 1.0% and 0.5% per annum respectively. Conversely, but also at a slow rate, coal and oil are projected to decrease at an average annual rate of –0.5% and –1.6%, respectively over the same period.

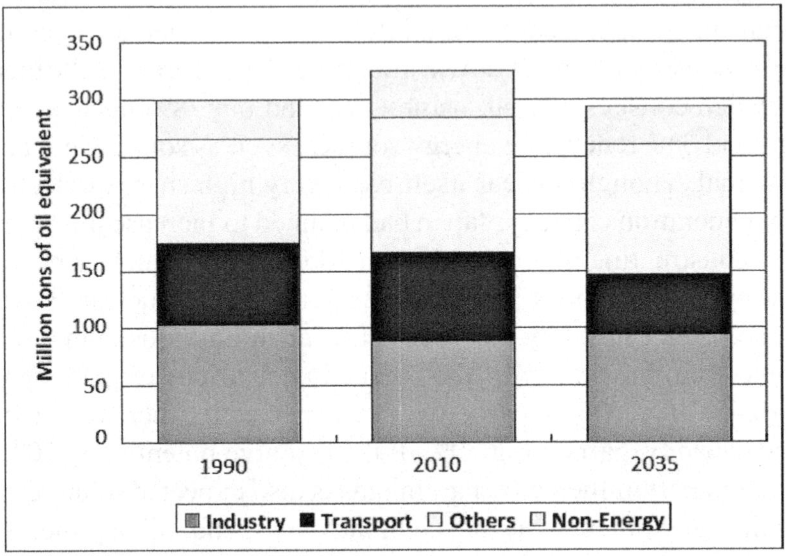

Figure 7A. Final Energy Demand by Sector.

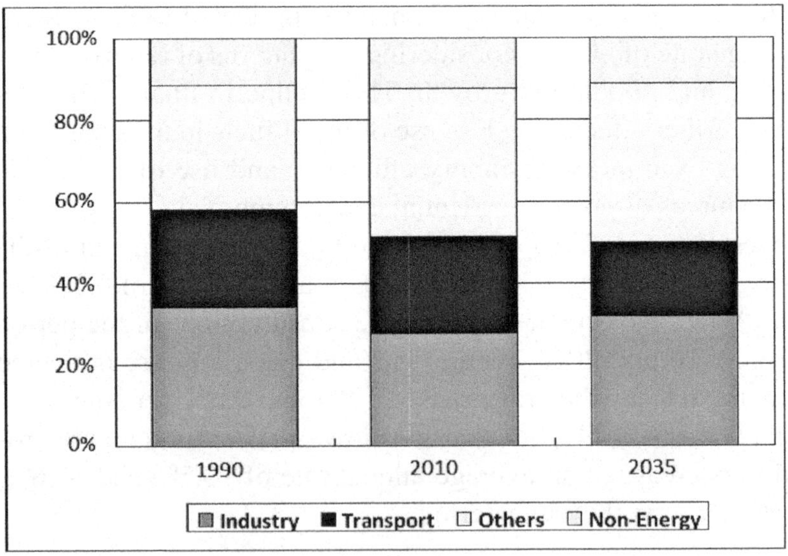

Figure 7B. Final Energy Demand by Sector.

The net primary energy supply from 2010 to 2035 is projected to decrease from 494.0 MTOE to 458.1 at an average annual rate of –0.3%. This is due to a decline in the supply of oil and nuclear, and it is expected to decrease at an average rate of –1.3% per annum for oil and –4.5% per annum for nuclear over the period between 2010 and 2035. Furthermore, the shares in total primary energy supply are projected to decrease from 38.7% to 31.7% for oil and 13.9% to 5.2% for nuclear over the same period. Nevertheless, over the same period, natural gas and coal will grow at an average rate of 1.6% and 0.3% per annum from 2010 to 2035.

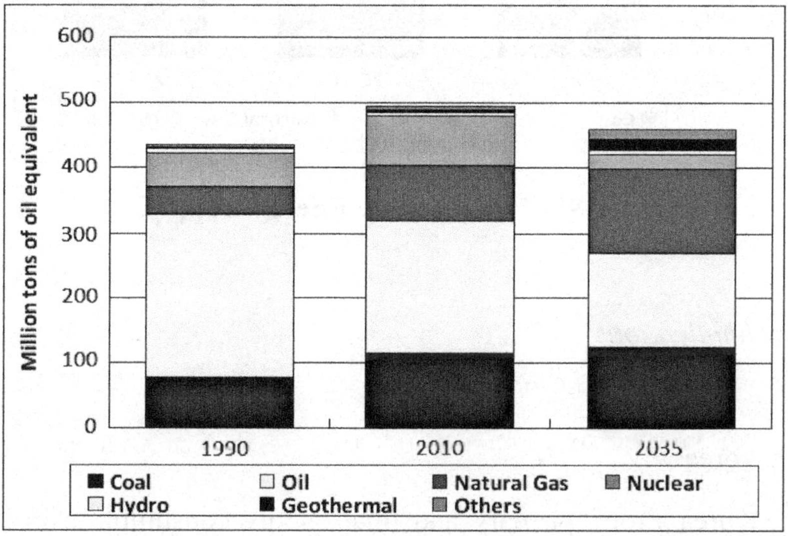

Figure 8A. Net Primary Energy Supply.

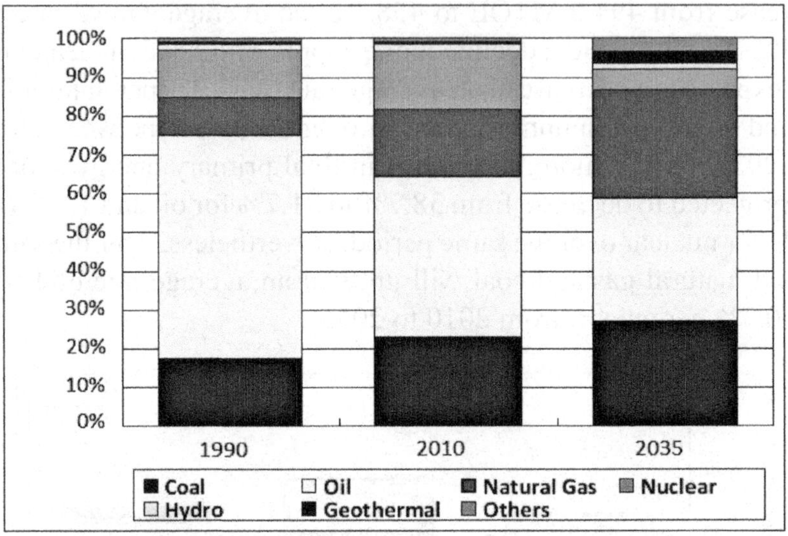

Figure 8B. Net Primary Energy Supply.

Policy Implications

South Korea

South Korea's total primary and final energy consumption in the 1990s had rapidly increased at a rate faster than that of GDP, whose growth has been driven by energy-intensive industries such as petrochemical, steel, and cement industries. Since 1997, the contribution of these industries to South Korea's GDP growth has gradually declined, resulting in reduced energy use. However, the shift to a less energy-intensive industrial structure takes time, which indicates that energy-intensive industry will prevail in the short to mid-term future. However, South Korea will and has to transform its industrial structure into a less energy-intensive one in the longer term.

Until now, South Korea has promoted the diversification of energy resources and suppliers to reduce excessive external energy dependence and to find new and renewable energy sources, which contributes to enhancing energy security as well as environmental preservation. It is highly recommended that South Korea continue with its current policy goals of transforming into a less energy-intensive and greener economic structure through implementation of policies to keep a balance between energy, the economy, and the environment. Such nationwide efforts and campaigns will eventually transform the South Korean economy into a less energy-intensive and greener one in terms of energy savings as well as reduced CO_2 emissions. Such an achievement would position South Korea as one of the world's leading nations in terms of low-carbon green growth.

China

As the world's largest developing country, China has been and will continue to maintain its fast GDP growth, which inevitably requires a large volume of energy to fuel this economic growth. While China's future energy demands will increase, the energy intensity (energy demand per GDP) will decline with the implementation of sound energy efficiency and non-fossil fuel technology policies. There is great potential for energy savings in China through a structural change in the economy with a shift in focus from heavy to lighter manufacturing and the development of China's services industry. The life span of China's buildings and infrastructure should be extended to prevent the rapid turnover and excessive production of energy-intensive products such as steel and cement.

It is recommended that small inefficient power plants, coal mines, and small energy-intensive industries like cement and steel plants should be closed to improve the industrial structure in China. In the longer term, energy efficiency in the residential, commercial, and transport sectors will be increasingly important in addressing energy saving given China's booming real estate and automobile markets in recent years. In addition, the development and deployment of alternative and renewable energy sources is needed to create a future environmentally-friendly energy market structure.

The Chinese government should establish and implement market-based measures to motivate enterprises and consumers to take immediate actions. More than anything else, energy pricing reform such as removal of the current energy subsidies, the establishment of energy taxes, and the establishment of a carbon tax should be initiated as soon as possible. As a more immediate action, China should draw on international experience to develop and implement Minimum Energy Performance Standards (MEPS) and energy efficiency labeling to ensure that industry and consumers are able to invest in high-efficiency technologies and appliances.

Japan

With net energy import dependence of about 82% and the majority of its oil imported from the Middle East, Japan has been trying to diversify energy sources as well as energy suppliers. Along with these efforts, Japan has been energetically implementing energy efficiency improvements and conservation policies and programs. Japan's energy intensity has been declining since 1980, reaching what is currently the lowest level in the world. This is largely ascribed to the enormous improvements in energy efficiency through technological innovation in Japan on both supply and demand sides. It should be kept in mind that Japan imports most of its energy sources and has inevitably been aggressive in improving energy efficiency to reduce its energy demand and to save its foreign exchange rates accordingly.

As a global leader in energy efficiency, Japan should share such successful policies with other countries at its earliest convenience. By doing this, Japan will be able to contribute to rational energy use in the global context, reducing world energy demand as well as CO_2 emissions. This would benefit the Japanese as well as international economies by increasing the amount of energy available on the market, especially in light of limited energy resources.

The current long-term projection over the period between 2010 and 2035 for energy demand and supply is based on the assumptions of previous administrations, namely the Ministry of Energy, Trade, and Industry's (METI) projection in 2010 and government discussion in 2012. The 4th Strategic Energy Plan was initiated and developed after undergoing a heated national debate.[4] New goals, policy directions, and strategies to attain it were proposed and approved under the Abe administration. Japan will minimize its dependency on nuclear power and maximize its use of natural gas and renewable energies as well, which is the starting point for rebuilding Japan's energy policy in the 2014 version of the Strategic Energy Plan.

Regional Energy Cooperation in Northeast Asia

Up until now, we have surveyed the current energy situation and energy future of three major countries in Northeast Asia: China, South Korea, and Japan. Given the differences in energy resource endowments and in current energy market structures among the countries in the region, NEA has a tremendous potential for greater regional energy cooperation. There are three big energy importers—China, South Korea, and Japan—and two big energy producers—China and Russia—in the NEA region. However, energy trades among the countries in NEA still remain at a meager level, in the sense that the energy importing countries in the region heavily depend on imports from other regions of the world, while the unique net energy exporting country in NEA, Russia, exports energy mostly to the European region.

Thus, regional energy cooperation among the countries in NEA has

4. Agency for Natural Resources and Energy, Ministry of Economy, Trade, and Industry, "Strategic Energy Plan" (Provisional Translation), April 2014, available from *www.enecho.meti.go.jp/en/category/others/basic_plan/pdf/4th_strategic_energy_plan.pdf*.

shown relatively slow progress compared with other regions of the world. Despite the huge potential for multilateral cooperation, most of the energy development cooperation projects in NEA have been planned and/or implemented at the unilateral or bilateral level without consideration of the regional market characteristics as a whole. Accordingly, many projects for energy development and trades in NEA faced uncertainty related to investment risk, future market size, and commercial feasibility and viability, so that they were delayed or could not be realized. There exist a number of dilemmas and barriers to promoting energy cooperation among the countries in the NEA region.[5]

Nevertheless, there is now a growing interest in NEA in promoting regional energy cooperation since the region has a great potential to improve its energy situation by developing those untapped energy resources and constructing cross-border energy transport infrastructures to facilitate the intra-energy trade between the countries in the region. Nevertheless, NEA has the same problems that other regions such as Southeast Asia, Europe, North America, and Africa are facing, which include energy security, energy efficiency improvement, and regional environmental conservation. Those problems have been resolved through multilateral cooperation at the regional level in these other regions of the world. This implies that other regions' energy cooperation experiences can be applied to the NEA region, too.

Need for Regional Energy Cooperation in Northeast Asia

Enhancing Energy Security. China, Japan, and South Korea are huge energy consuming and importing countries. They are faced with ever-lasting challenges regarding energy security. Thus, major interest in promoting energy cooperation in NEA originally came from the countries' needs for expanding their energy security ca-

5. Chang-Won Park and Ji-Chul Ryu, *Energy Cooperation in the NEA: Long-term Visions and Strategy*, Seoul: Korea Energy Economics Institute and the Ministry of Trade, Industry, and Energy, March 2005.

pability by expanding energy trades with neighboring countries through promoting regional energy cooperation.

In NEA, there exists a great potential for oil and gas development in the Russian Far East as well as for coal in China and Mongolia. However, these resources are located in remote areas and large investments are required for their development. As capitol-rich countries, China, Japan, and South Korea can participate in the energy development in energy resource-rich countries such as Mongolia and Russia. They have increasingly shown their interest in developing huge untapped energy resource reserves in remote areas in neighboring countries. Development of such huge energy reserves in NEA is, if accomplished, expected not only to bring a fundamental change in the regional energy patterns but also to significantly affect the international energy market.[6]

Also, given their geographic proximity, the countries in NEA can gain many benefits by promoting energy trade with neighboring countries. For example, while it takes 20 days to transport oil from the Middle East to NEA, bringing supplies from Sakhalin facilities in Russia only requires three to five days. Therefore, greater intraregional trade within the NEA would significantly improve transport time and potentially also improve transport security as supply lines are shortened. Trading natural gas and/or electricity from neighboring countries through trans-boundary energy grid networks also can contribute to relieving energy shortage problems in the region.

Development of Energy Trade Infrastructures in Northeast Asia. In order to facilitate intra-regional energy trade, construction of cross-border energy supply infrastructure—in particular natural gas pipelines and power transmission networks—will be required in NEA. The large energy-importing countries—China, Japan, and South Korea—are expected to gain many benefits in terms of eco-

6. Ji-Chul Ryu, *Towards the Establishment of Institutionalized Framework for Multilateral Energy Cooperation in the NEA*, Seoul: Korea Energy Economics Institute and the Ministry of Trade, Industry and Energy, August 2003.

nomics and market efficiency, since they will be able to trade energy with the neighboring countries if the countries are geographically connected with an energy grid network.

Also, countries in NEA can share the common benefits from the establishment of cross-border energy trade infrastructures between the countries, allowing them to proactively respond to a number of challenges, including managing peak electricity demand in the summer, addressing emergencies such as energy supply crises, advancing the effective exploration and utilization of regional energy resources, and improving the stability of long-term transport by sea. As such, a greater number of regional energy grid networks have been suggested as a method to facilitate cheaper and more optimal deployment of resources among the countries in the region and for securing markets for exports. At this time, South Korea has to run the risk of building an energy grid through North Korea.

Contribution to the Establishment of a Peace Mechanism in Northeast Asia. Increasing interdependence among countries in the NEA region in terms of economic relations, trade, and investment is one of the key factors for the improvement of political stability in the region, as observed in Europe, North America, and Latin America. Facilitating energy trades among the countries in NEA, particularly by constructing cross-border energy infrastructures, is also expected to contribute to the alleviation of political tension in NEA as well as in the Korean peninsula, which has been escalating these years. Connecting the energy infrastructures between South Korea and Russia or between South Korea and China will call for the involvement of the Democratic People's Republic of Korea (DPRK, or North Korea) due to geography, which will, in turn, play an important role in resolving the chronic energy shortage problems in the DPRK.

Eliminating the Asian Price Premium. Countries in Northeast Asia, namely China, Japan, and South Korea, pay a higher price for crude oil imported from the Middle East, which is the so-called "Asian Price Premium." Also, the price gap between the gas markets of Asia and those of North America and Europe has significantly widened because of tightened gas markets in Asia. Thus, it is expected that an increased supply of non-Middle Eastern oil, including oil from Eastern Siberia and the Pacific Ocean, will contribute to easing the region's premium on oil prices to a considerable extent.

Governments and research institutions in China, Japan, and South Korea have discussed relevant action plans for easing the Asian Price Premium, while China and Japan have also individually formed and managed bilateral and multilateral frameworks for cooperation with other countries.[7] To further these efforts, it is time for countries in NEA to work together to jointly develop a new pricing mechanism that reflects the current gas market situation and that offers more flexible conditions for gas imports.

Establishing a Regional Common Energy Market. As NEA seeks opportunities to strengthen regional energy cooperation, the establishment of an integrated energy market in the region could yield significant energy and economic benefits. Such benefits include providing for security of energy supply and demand; decreasing transaction costs in regional trade; promoting greater efficiency of scale in the energy sector; offering access to affordable modern energy; and reducing emissions from pollutants typically associated with energy production and use as countries have access to a greater range of lower-carbon sources.

7. On June 7, 2013, Japan's Ministry of Economy, Trade and Industry and the European Commission's Directorate-General for Energy decided to launch cooperation on gas and LNG market, expecting that it will contribute to promoting the development of transparent and liquid global gas market driven by supply and demand fundamentals. The two parties agreed to discuss sustainable ways to improve oil-indexed gas prices used in European and Asian markets. In the case of China, its state-owned CNPC has continued discussion and joint research with companies from South Korea and Japan to deal with the Asian Price Premium.

Enhancing Complementaries among the Countries through Energy Cooperation.[8]

Countries in NEA each have their own advantages in energy that are highly complementary and have many common interests. The benefits of cooperation far outweigh those of competition. China is rich in coal resources and has advantages in geographic location, oil and gas exploration technology, and energy production. It also has a high demand for advanced new energy and energy-saving technology. South Korea has advanced science, technology, and management techniques, has considerable experience with oil reserves and good refining abilities, and has much successful experience in energy saving and market operation. Russia is rich in coal, oil, and gas resources and has put emphasis on developing its East Siberia and Far East regions, as well as actively expanding into the Asia-Pacific market in recent years, including the construction of a main pipeline from East Siberia to Pacific Ocean (ESPO) to bring crude oil from Russia into the energy consumption market of NEA. In addition, countries in the NEA have obvious complementary aspects regarding energy production, processing, and transportation. Enhancing regional cooperation may produce advantages for each region and allow them to learn from each other, thus producing a win-win situation.

Harmonization of Energy Standards. There will be great benefits gained through international harmonization of technological and safety standards, which will facilitate public-private energy projects in the NEA region. Applying advanced technologies to modernize existing energy infrastructure or building new facilities provides benefits for national energy independence as well as regional environmental preservation. Currently, national variations in safety regulations and technological standards present an obstacle to internationally standardized project designs, which would aid the growth

8. In November 2005, the United Nations Economic and Social Commission for Asia and the Pacific (ESCAP), established the Intergovernmental Collaborative Mechanism on Energy Cooperation in North-East Asia, see ESCAP, "Annual Report: 19 May 2005 - 12 April 2006," E/2006/39, E/ESCAP/1390, Bangkok: United Nations, 2006.

of all economies involved. Achieving the harmonization of industrial technology and safety standards could overcome this obstacle, facilitating the emergence of a regional market that offers a choice of mutually-beneficial projects and positively influencing business-to-business cooperation across countries in the NEA region. Each NEA country has its own approach to energy prices influencing the energy mix, emphasis on the role of nuclear energy, policies aimed at reducing greenhouse gas emissions, national taxation policies, environmental standards, investment frameworks, and technological progress. For a balanced NEA regional energy market to develop, ideally the countries need to work on introducing harmonized commercial, regulatory, and political measures that create incentives for ensuring sufficient energy security for all.[9]

Opportunity of Regional Energy Cooperation in Northeast Asia

Overview. In recent years, a number of discussions and negotiations have occurred between energy producing and exporting countries (e.g. Russia and Mongolia) and energy consuming and importing countries (e.g. South Korea, China, and Japan); most of these have taken place on a bilateral level but some have also involved multilateral governmental and industrial cooperation. These cooperation projects can be broken into roughly four categories:

- Enhancing cooperation in upstream, midstream, and downstream for the oil, gas, coal and electricity sectors, including building cross-border energy supply infrastructure, such as pipelines and interconnected power networks to facilitate intra-regional energy trade in NEA;
- Cooperation for energy conservation;
- Cooperation for nuclear power generation safety; and

9. ESCAP, "Five-Year Strategy (2010-2014) to Implement the Inter-governmental Collaborative Mechanism on Energy Cooperation in North-East Asia," Bangkok: United Nations, January 2010.

- Promoting policy cooperation at the multilateral level, including information sharing, facilitating policy dialogues, and enhancing mutual understanding in the energy sector.

Energy and Infrastructure Development Cooperation. There exists great potential for oil and gas exploration and development in the Russian Far East as well as for coal in China and Mongolia. China, South Korea, and Japan have shown increasing interest in developing the huge, untapped energy resource reserves in remote areas of neighboring countries. Development of such huge energy reserves in NEA is, if accomplished and penetrated in the huge demand market in the region, expected not only to fundamentally change the regional energy patterns, but also to significantly affect the international energy market.

Table 1 highlights cooperation projects that have occurred or been proposed in NEA that are related to a range of energy source exploration and development in the upstream sector, construction of energy grid networks in the midstream sector, including more specific efforts to target coal, oil, gas, electricity, and other energy resources. Some midstream projects to construct pipelines and power grid networks have already been undertaken on a bilateral basis but need to be implemented at the multilateral level if they are to penetrate multiple countries in the region. Railway systems also need to be included in the coal midstream activity, as it is an important transport mode for coal. More details on cooperation activities among the countries in the NEA region will be discussed in the following section.[10]

All the countries in the NEA region are actually implementing policies for energy conservation/efficiency and renewable energy development in pursuit of a sustainable future energy system and to achieve green economic growth. Development of renewable energy resources, saving energy, and improving the efficiency of energy use

10. Sungkyu Lee and Ahhyun Park, "Challenges and Tasks of Energy Cooperation in the NEA," paper presented at the Pacific Energy Summit, Seoul, Republic of Korea, June 2014.

can greatly save on investment in energy development, reduce energy consumption, and improve economic efficiency, which is, in turn, an effective way to guarantee energy security.

Countries in NEA need to develop a mechanism for cooperation that actively facilitates energy saving and comprehensive utilization of renewable energy sources, so as to carry out project cooperation in renewable energy development/deployment and the improvement of energy efficiency. South Korea and Japan are recognized world leaders in having high energy efficient economies and effective dissemination of their best practices should bring positive results in many energy efficiency measures, which China, Mongolia, or Russia could undertake. Appropriate technology and information exchange mechanisms need to be established in NEA.

Cooperation for Energy Conservation/Efficiency Improvement and Renewable Energy. All the countries in the NEA region are actually implementing policies for energy conservation/efficiency and renewable energy development in pursuit of a sustainable future energy system and to achieve green economic growth. Development of renewable energy resources, saving energy, and improving the efficiency of energy use can greatly save on investment in energy development, reduce energy consumption, and improve economic efficiency, which is, in turn, an effective way to guarantee energy security.

Countries in NEA need to develop a mechanism for cooperation that actively facilitates energy saving and comprehensive utilization of renewable energy sources, so as to carry out project cooperation in renewable energy development/deployment and the improvement of energy efficiency. South Korea and Japan are recognized world leaders in having high energy efficient economies and effective dissemination of their best practices should bring positive results in many energy efficiency measures, which China, Mongolia, or Russia could undertake. Appropriate technology and information exchange mechanisms need to be established in NEA.

Cooperation for Nuclear Power Generation Safety. In the wake of the Fukushima nuclear accident that occurred in Japan in March 2011, there have been growing concerns worldwide on the safety of using nuclear power. Also, as observed from the nuclear power plant accident at Chernobyl (April 1986) in the former Soviet Union, the ramifications of a large-scale nuclear accident have serious effects that reach beyond a state's borders into neighboring countries. Given the vulnerability of nuclear facilities to natural disasters in Japan and also the ambitious expansion of nuclear power plants in China, in particular, the need for regional coordination for the safety of nuclear power plants in NEA has significantly increased. Thus, multilateral coordination and cooperation to ensure safety of nuclear power plants should be a regional energy cooperation concern in NEA.

Accelerating Multilateral Approach for Regional Energy Cooperation. The construction and operation of cross-border energy transportation networks requires multilateral cooperation if they involve several countries in the project. By taking a bilateral approach, some problems may be unsolvable, such as natural gas pipelines involving a third country and trans-national transportation. If the problem is solved by building a detour, costs may rise and the economic efficiency may decline. On the other hand, if productivity and demand cannot reach a balance, or if there is a production loss or supply shortage problem, this would result in economic losses to both supplier and consumer. Thus, it would be rational to promote energy cooperation in NEA to build an interconnected system among the countries. This would require the countries to conduct policy coordination and cooperation on investments and would require the governments of the countries to participate directly, so as to enhance inter-government coordination capacities, improve the levels and standards of multilateral, long-term and large-scale energy cooperation, consolidate and develop the material foundation of NEA cooperation, strengthen the international mutual trust, facilitate the economic development, and political stability, accelerate progress on regional integration, and achieve positive regional economic effects in Northeast Asia.

Multilateral energy cooperation may help the production and economic sectors in NEA form an optimal portfolio, producing favorable energy economic benefits. Energy cooperation involves energy resources, capital, labor, transportation routes, machinery, relevant technology and industrial development, environmental protection and other productive factors. These factors are currently unevenly distributed and vary in quality among the countries of Northeast Asia. It is only through multilateral cooperation that the optimal portfolio of these factors in the energy development process can be realized.

Cooperative Project by Energy Sources in Northeast Asia		
Sector	Petroleum	Natural Gas
Up-stream	• Joint oil development in the Russian Far East/ Equity participation	• Joint gas development in the Russian Far East/ Equity participation • Unconventional gas development
Mid-Stream	• China-Russia oil pipeline connection /construction • Import of Russian oil through pipeline and oil tankers • Commercial use of the North Pole Route	• Building of gas transport network among the NEA countries (China-Russia, Japan-Russia, ROK-China,. ROK-Japan, Russia-China-ROK, Russia-DPRK-ROK) • Gas pipeline connection (ROK, DPRK & Russia) • Participation in liquefaction facility construction in Russia • Import of Russian gas via pipeline and LNG carrier • Commercial use of the North Pole Route
Down-Stream	• China-Russia refinery construction / Petroleum product trades	• Enter the market of city gas business

Cooperative Project by Energy Sources in NEA, Continued	
Coal	Electricity & Renewables
• Joint coal development in the Russian Far East • Joint coal development of South Gobi desert in Mongolia	• Joint construction of electricity supply facility (power plant and transmission grid) in the Russian Far East • Joint renewable energy development
• Construction of railway to transport mineral resource of Mongolia • Expansion/Repair & maintenance of DPRK's transport facilities • Railway connection between Russia and two Koreas. • Railway connection between China & Mongolia • Import Russian and Mongolian coal through railway and barge	• Establishment of Asia Super Grid (China-Russia, Japan-Russia, Russia-Mongolia, ROK-China, ROK-Japan, China-Mongolia) • Power grid connection among the two Koreas and Russia • Import of Russian electricity
• Clean coal, coal to liquids technology cooperation	• Power end-use energy efficiency

Table 1.[11]

11. Ibid.

Major Issues in Energy Cooperation with Countries in Northeast Asia

Regional energy cooperation among the countries in the region has shown relatively slow progress as compared with other regions of the world. This has been a result of various factors, including high capital requirements for energy and infrastructure development; the lack of infrastructure, resource nationalism in energy-rich countries; capital accessibility problems for energy development projects; uncertainty surrounding the development of energy projects in Russia; the geopolitics of the NEA region; market inertia in the existing energy industry in each country; lack of market compatibility and multilateral mechanisms for regional energy cooperation; intensified competition for energy security; and conflict between multilateral and bilateral relationships. This section briefly reviews those dilemmas and barriers hindering the promotion of energy cooperation among the countries in the region.

High Capital Investment Requirements, Lack of Infrastructure, and Resource Nationalism. Energy development projects usually require huge capital investment, and are often characterized by a high risk and slow return. Also, constructing energy production and delivery facilities, such as power plants and pipelines, is very capital-intensive and requires a high level of technology application. Therefore, these energy projects should always be implemented subject to the procurement of capital and technology as required, and also subject to securing demand markets to guarantee investment return.

In the NEA region, most oil, natural gas, and coal reserves are located in remote areas isolated from the regions of demand. The development of those energy resources in NEA as well as the construction of transport infrastructure requires relatively higher capital investment compared to that in other regions. Northeast Asia also suffers from the lack of infrastructure necessary for energy development and trans-boundary energy transportation.

However, energy-rich countries in NEA, namely Mongolia and Russia, are not capital-rich. The financial capability of Russian energy companies appears to have significantly deteriorated since the late 2000's, when the global financial crisis of 2008 caused an economic downturn and energy prices declined, affecting Russia's export market in Europe. The Mongolian economy experienced a similar situation due to declined resource prices since then. Thus, there has been growing concern and doubt that these countries would be able to finance energy development projects alone.[12]

In the case of Mongolia, the government implemented a policy to attract foreign direct investment (FDI) in order to overcome these obstacles and commence energy projects, particularly the development of its large coal mines in the Tavan Tolgoi fields, which has great potential to become a major exporter of coking coal to other NEA countries. However, the coal mines in Mongolia have the serious problem of a lack of infrastructure, such as roads, railways, and power supplies, which will require massive foreign investment to develop the large-scale infrastructure.

Russia also needs massive investment for energy and infrastructure development in Eastern Siberia and the Russian Far East. However, Russia has shown a tendency toward so-called "resource nationalism." There has been increasing government influence on the ownership and development of energy resources in its territory for strategic and economic reasons through state-owned energy companies, like Gazprom and Rosneft, particularly since the 2000s. This tendency of the Russian government has been shown to block foreign capital investment in attractive energy development projects in Russian territory, and thus Russia's energy supply potential to NEA countries has not yet been enhanced. This factor has also weakened Russia's competitiveness against other regions, for example, Australia, the Middle East, and the Americas as an energy supplier to the

12. See National Reports from China, Korea, Mongolia, and Russia in Greater Tumen Initiative, "Capacity Building Training Project on Energy: Policies and Prospects for Regional Energy Cooperation," Beijing, 2014.

NEA countries.

Project Uncertainty. The future potential of increased energy trade of oil and natural gas and of enhanced regional energy cooperation will eventually depend on how much Russia will increase energy production and supply from Eastern Siberia, the Russian Far East, and Sakhalin Island, and also on how soon or how fast cross-border energy transport infrastructure can be established in the NEA region.[13] According to Russia's plan for the ESPO pipelines, Russia will supply 80 million tons of oil per annum to the Asia-Pacific region through the pipelines. However, several major uncertainties lie ahead for the realization of the oil supply plan in the Russian Far East:

- The quantity of the oil resources needed to fill up the trunk pipeline is unknown;
- Oil production in underdeveloped oil and gas fields in Eastern Siberia and the Sakha Republic is not promising yet; and
- Exceptionally high price of oil through this pipeline due to high construction costs of U.S. $12 to 16 billion for the project plus high transport fees.

Considering the above-mentioned obstacles to the oil supply reaching the pipeline, it is quite difficult to predict Russia's potential to supply a significant amount of oil to the NEA region from East Siberia.

Sakhalin Island is regarded as a strategically important oil and gas import source for the NEA countries because of its geographical proximity, and it is expected that oil production from Sakhalin will be increased significantly in the future. Russia also recognizes Sakhalin as a strategic region that would expand the export market,

13. Ji-Chul Ryu, "Energy Cooperation between the Two Koreas," Seoul: Korea Energy Economics Institute and Ministry of Commerce, Industry and Energy, June 2005.

which is currently dependent on Europe, to Asia-Pacific countries. However, there are some difficulties in Sakhalin oil and gas development because of the presence of sea ice in the winter and environmental protections. Offshore exploration costs in Sakhalin have increased and these trends are expected to continue. The price of crude oil imported from Sakhalin was shown to be higher than that from the Middle East.[14]

Thus, Russia's potential to supply gas to the NEA countries depends largely on the speed of implementation of the Unified Gas Supply System (UGSS) Eastern Program and the ESPO Pipeline Plan. The possibility of importing Russian energy sources will be a positive option in the long-term perspective, but it is still uncertain in the short-term. A key question for Russian energy is how attractive and reliable a supplier Russia will be in the future, compared with other sources in the world.

Historically, NEA countries have imported their energy sources mainly from the Middle East, Australia, and Southeast Asia. Russia's hesitation in developing energy resources and transport infrastructure to the NEA region will make the largest energy importing countries in the region, China, Japan and South Korea, move from Russia to more reliable and attractive export sources. Then, Russia will lose the market, and the growth of regional energy cooperation in NEA through expanding intra-regional energy trade is likely to be delayed.

Political Instability. Political stability is an important factor in energy projects, particularly in relation to the security of projects. There still exists a serious bilateral territorial dispute between Japan and Russia over four of the Kurile Islands, and more seriously, political instability still remains on the Korean Peninsula. Traces of the Cold War still remain, most notably on the Korean Peninsula. Uncertainty regarding political stability involves a high risk for investors on

14. When the import price of Vityaz crude oil from Sakhalin is compared with Oman crude oil, which is similar in American Petroleum Institute (API) gravity and sulfur, the import price of Vityaz is higher than that of Oman.

energy projects. Routing the pipeline network from Russia to South Korea through the DPRK would probably be one of the most difficult problems unless a political agreement between two Koreas is settled.

Lack of Market Compatibility Among the Countries in the NEA Region. Countries in the NEA region have a diversity of market systems and structures, capacities of financing investment, human resources, and energy policy/planning, which are regarded as major factors that could impede the facilitation of regional energy cooperation in NEA. They lack the basic framework for economic cooperation between the market and non-market systems; the lack of mutual trust, especially between two systems. Countries in the region will need to make concerted efforts to narrow such gaps by coordinating their policy directions and policy tools.

Lack of Multilateral Mechanism for Regional Energy Cooperation in NEA. At present, there is no concrete framework for formalizing and institutionalizing multilateral economic and energy cooperation among the countries in NEA. Most economic relations among countries in the region are under development on a bilateral basis. Similarly, most of the energy projects undertaken by countries in NEA have been planned or implemented unilaterally or bilaterally. As a result, many projects for energy development, production, transportation, and trade in the region have faced unnecessary, additional uncertainties with regard to financing, securing an export market, and addressing regional security issues.[15] This complicated situation makes it harder than ever for solutions to materialize and to promote projects and trade. In addition, it also implies that financing for energy projects significantly depend on a government's financial support or a private company's own funds, rather than drawing on regional or global sources.

Intensified Competition for Energy Security. Major economic powers in the NEA region have been competing against each other to secure energy resources. In particular, China and Japan have

15. Ji-Chul Ryu, "Energy Cooperation with the DPR Korea in the International and Regional Context," Seoul: Korea Energy Economics Institute, December 2002.

been engaged in a fierce competition to secure energy supplies. Their energy-related efforts have included conducting negotiations with Russia and offering to provide Moscow with financial incentives for the installation of fuel pipelines in Russian territories to benefit their respective economies. Energy-importing countries in the region are busily devising and implementing national energy strategies that are intended to reinforce energy security, without considering common regional concerns. A typical example was the rechanneling of the ESPO oil pipelines in Russia. Originally, Russia proposed to build the ESPO to China in 1994; however, the plan has been delayed and modified many times by Japanese intervention since then.

Conflict Between Multilateral and Bilateral Relationships. Given the absence of concrete institutional arrangements for regional energy cooperation at the multilateral level, most energy cooperative projects in the region are implemented on a bilateral basis. As previously discussed, strong bilateral energy cooperation activities have taken place, particularly between China and Russia, in the oil, natural gas, and electricity sectors. China also has aggressively promoted bilateral cooperation with Mongolia and the DPRK for coal development and infrastructure construction. Meanwhile, South Korea and Japan seemed to play minor roles in promoting regional energy cooperation in the NEA region, although they have a great capability to participate in energy and infrastructure development projects and would gain some significant benefits from the cooperation. Russia, although it recognized a strong necessity for regional energy cooperation in NEA, appears to be passive in promoting multilateral energy cooperation in the region but has been more inclined toward strengthening bilateral relationship with individual countries in NEA for strategic reasons.

Such strong initiatives by some countries toward strengthening bilateral energy cooperation need to be shifted toward facilitating regional energy cooperation at the multilateral level in pursuit of common energy goals and economic prosperity in the NEA region.

Energy Grids in NEA

Oil Pipelines. Russia established an oil pipeline system in the NEA region, called the Eastern Siberia-Pacific Ocean (ESPO) pipeline, with a branch pipeline to Daqing, China, which created an outlet for East Siberian oil to get to the Asia-Pacific market. The first phase of the ESPO pipeline was completed in December 2009, stretching a distance of 2,757 km from Taishet to Skovorodino. The second stage was completed in 2012 and runs a distance of 2,100 km from Skovorodino to Kozmino. A total of 400,000 billion barrels per day (bbl/d) of crude oil was sent through the pipeline in its first year of operation. Also, the 997 km Russia-China oil pipeline was constructed from Skovordino to Daqing in September 2010. The amount of oil to be delivered through the ESPO pipeline to the port of Kosmino is expected to be 80 million tons per annum by 2020, and the Skovorodino-Daqing branch pipeline has a capacity of up to 25-30 million tons of oil per annum.

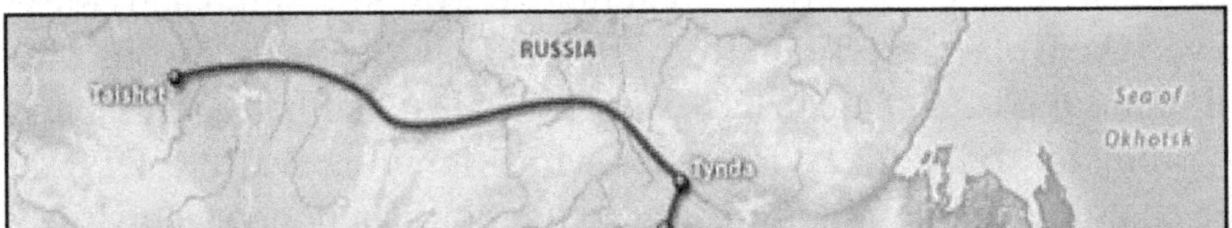

Figure 9. East Siberia-Pacific Oil Pipeline.

Rosneft, a state-run oil company in Russia, emerged as Russia's top producer recently and conducts operations, oil development, and production business in the region. Another state-run company, Transneft, runs the ESPO pipeline system.

Natural Gas and Infrastructure Development. The Russian Far East (Irkutsk, Yakutsk, and Sakhalin Island) contains one of the largest unexplored natural gas bearing areas still left on earth. The area is a potential large source of natural gas for markets in South Korea, China, and Japan. The development plan for natural gas in NEA also includes a project for the construction of cross-border pipeline networks as a part of an existing gas export line from gas fields in Eastern Siberia to the NEA market. The realization of these projects will provide countries in the region with diverse energy sources (in addition to coal and nuclear) while possibly improving environmentally-friendly energy systems. Eventually, the project will contribute to the creation of an integrated regional energy system in NEA.

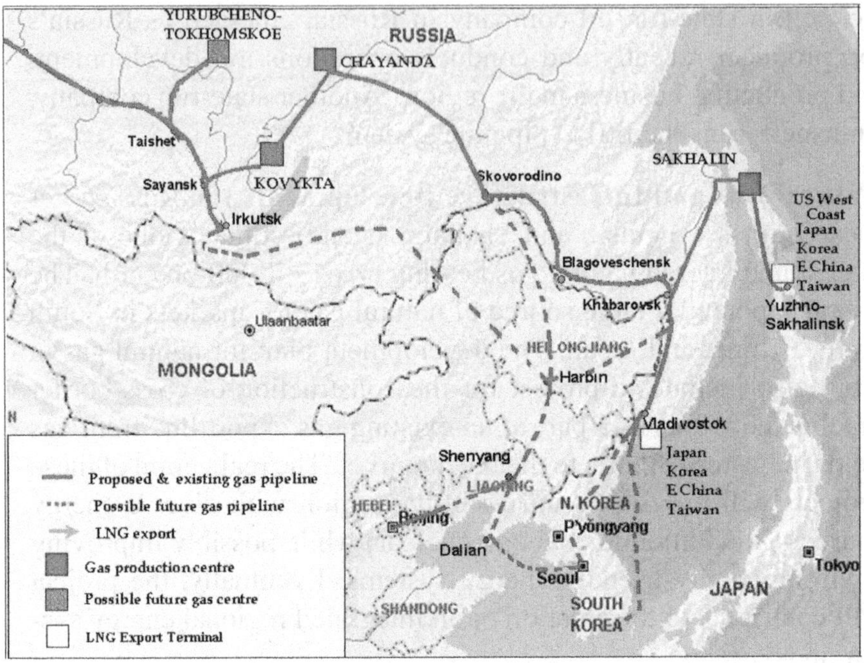

Figure 10. Gazprom's UGSS Eastern Program.

According to Gazprom's plan on the UGSS Eastern Program, four gas production centers will be developed to supply natural gas to the NEA and Asia-Pacific regions, which include:

- Krasnoyarsk center, based around Sobinsko-Paiginsk and Yurubcheno-Tokhomskoye deposits,
- Irkutsk center, based around the Kovykta field and fields in the north of the Irkutsk region,
- Yakutsk center, based on the Chayandinskoye field and neighboring oil and gas fields, and
- Sakhalin center, based on deposits on the continental shelf zone around the island.

Projected gas production in Eastern Siberia and the Russian Far East is assumed to be in the ranges of 44.9 – 88.2 billion cubic meters (bcm) per annum in 2015 and 76.6 – 133.8 bcm per annum in 2020, and to maintain constant production until 2034. Construction work on the Yakutsk gas production center, based on Chayandinskoye condensate field, is ongoing and is expected to be finished in 2017. Natural gas production at the Yakutsk center is expected to be 25 bcm per annum in 2021.

According to Gazprom's plan, a natural gas pipeline will be constructed and connected systematically from East Siberia and Sakhalin to Vladivostok by 2020, supplying both the domestic market in Russia and providing exports to neighboring countries via its trunk pipeline system (UGSS) as new gas fields are developed in the future. Pipeline development includes:

- First stage: 1) Constructing a gas pipeline in the Sakhalin-Khabarovsk section, 2) Building a trunk line and supplying gas to the Irkutsk area;
- Second stage: Development of Chayandinsk gas field 1) constructing a branch line to supply gas to China from Blagoveshchensk, 2) connecting the gas trunk line between Irkutsk and Taishet, and 3) connecting trunk line between Khabarovsk and Vladivostok and supplying natural gas to South Korea;
- Third stage: Developing fields in the Krasnoyarsk region and connecting line to the Taishet region;
- Fourth stage: 1) connecting gas fields between Irkutsk and Chayandinsk, 2) connecting gas trunk line in Blagoveshchensk and Khabarovsk section (completion of UGSS).

The Yakutia-Khabarovsk-Vladivostok gas pipeline, which will run for a considerable distance along the same corridor as the ESPO oil pipelines, will connect with the Sakhalin-Khabarovsk-Vladivostok pipeline in the future to form the Russian Far East gas pipeline infrastructure system. The next step will be a full-scale development of the Kovykta field in the Irkutsk region and the construction of a gas pipeline between Irkutsk and Yakutia gas production centers (about 800 km). As a result, a full-scale gas delivery system on Russia's Pacific coast will be constructed.

Supplying Russian natural gas to the NEA market will provide good energy security, because it will rely on the gas produced from large reserves nearby: First from the Sakhalin shelf, then from Yakutia, and in the long term from Eastern Siberia. In order to facilitate the export of Russian gas to the NEA and Asia-Pacific markets, an LNG plant and export terminal are also planned for construction at Vladivostok.

In May 2014, Russia's Gazprom and the China National Petroleum Company (CNPC) signed a 30-year sale and purchase contract for the China-Russia East Gas Line. As stipulated in the contract, Russia will begin to supply natural gas to China in 2018, with the gas supply increasing each year until it reaches 38 bcm per annum.

South Korea hopes to import at least 10 bcm of the Russian natural gas by 2017 at the earliest, in addition to the 1.5 million tons of LNG imported from the Sakhalin II project since 2008. This is also what the Russian side indicated during the signing of the intergovernmental agreement for natural gas cooperation between South Korea and Russia in October 2006. The Korea Gas Corporation (KOGAS) and Gazprom have had several consultative meetings since then on how to achieve this goal. Meaningful progress was made at the summit meeting between South Korea and Russia in Moscow, September 2008. South Korean President Lee Myung-bak and Russian President Dmitry Medvedev agreed to jointly undertake a U.S. $100 billion project that will include the development of natural gas pipeline routes from Siberia to South Korea. It features a natural gas pipeline running through the divided Korean Peninsula. The proposal calls for South Korea to import 10 bcm of gas a year for 30 years by 2015. South Korea plans to buy nearly 20 percent of its future natural gas from Russia through a pipeline that would pass through North Korea.[16]

16. National Reports from China, Korea, Mongolia, and Russia in Greater Tumen Initiative, "Capacity Building Training Project on Energy: Policies and Prospects

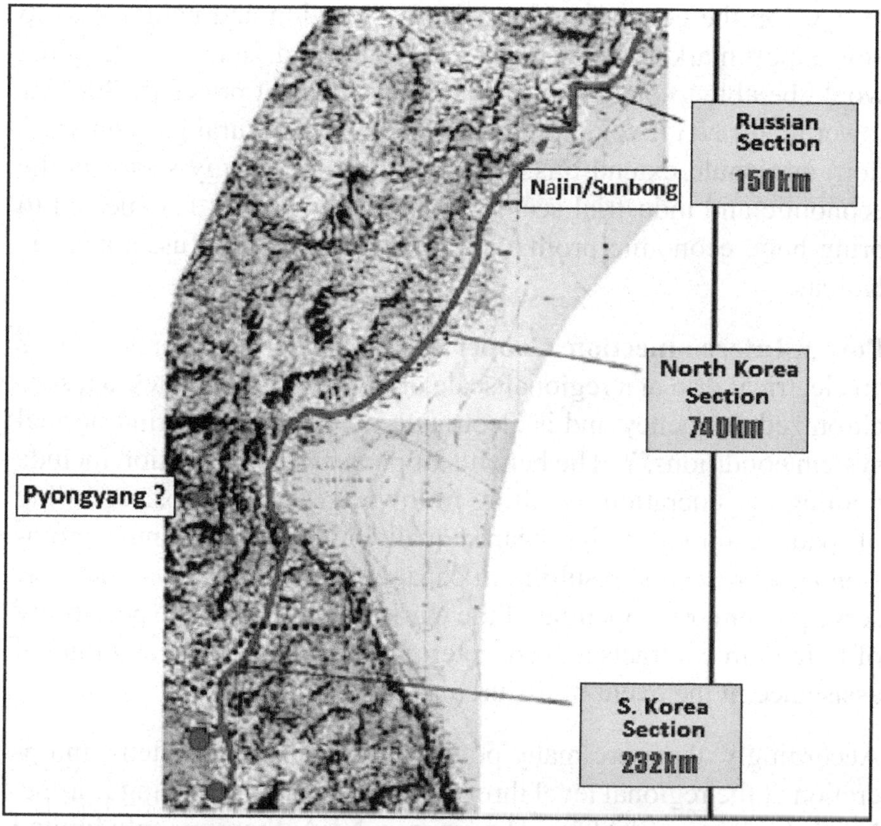

Figure 11. Russia - Two Koreas Natural Gas Pipeline.

To realize the gas pipeline from Russia to South Korea through DPRK territory it is absolutely necessary for the two Koreas to cooperate with each other closely. If the gas pipeline project is completed successfully, both countries would benefit economically. North Korea can earn annual transit fees of U.S. $150 million, while also profiting from the labor costs and regional development by participating in pipeline construction. The DPRK would also be able to alleviate its own power shortage problems by building gas-generated power plants in areas that the pipeline passes through. Russia could earn profits exceeding U.S. $90 billion by procuring a

for Regional Energy Cooperation."

stable gas market for 30 years and may extend this project in order to develop the economy of the Far East region and gain access to new export markets in South Korea, China, and Japan. South Korea would be able to obtain a stable supply of gas at prices by 30-70% lower than LNG (excluding the costs of piped natural gas construction) and could extend this cooperation in the energy sector to the economic and industrial sectors.[17] Thus, this project is expected to bring huge economic profits and other benefits for Russia and the Koreas.

Power Interconnection Cooperation. "Power interconnection is an electrical grid at a regional scale or greater that operates at a synchronized frequency and is electrically tied together during normal system conditions."[18] The benefits of power interconnection include pooling of generation, resulting in lower generation costs; pooling of load, resulting in significant equalizing effects; common provisioning of reserves, resulting in cheaper primary and secondary reserve power costs; opening of the market, resulting in the possibility of long term contracts and short term power exchanges; and mutual assistance in the event of disturbances.

Accordingly, there are many power interconnection systems in operation at the regional level throughout the world, including Europe, North America, and Africa. However, in NEA, there are only limited cross-border power interconnection systems. One has been established between Russia and Mongolia, and recently China and Russia began to establish a system between their two countries. There are also some proposals and studies for the establishment of power interconnection between three countries in NEA, namely Russia, North Korea, and South Korea. More ambitiously, the Asian Super Grid System is also under review among the NEA countries.

---

17. Youn Seek Lee, "The Gas Pipeline connecting South Korea, North Korea, and Russia: Effects, Points of Contention, and Tasks," KINU Policy Study 11-05, Seoul, Korea Institute for National Unification, 2011.

18. Wikipedia, "Wide area synchronous grid," available from *en.wikipedia.org/wiki/Wide_area_synchronous_grid*, last modified December 31, 2014.

Currently, there is an existing but weak connection between electrical power systems in East Siberia and the Russian Far East with a capacity of 220 kilovolts (kV), between Russia and Mongolia with a capacity of 110.220 kV and length of 380 km, as well as between Russia and China with a capacity of 110, 220, and 500 kV and a length of 206 km. Through these transmission lines, Russia exported 983TWh to China and 206TWh of electricity to Mongolia in 2010.

As for the power interconnection project between Russia, the DPRK, and South Korea, the project was commenced with the signing of a memorandum between Skolkovo Tech, En+Group, and the Korea Electricity Power Corporation (KEPCO) during the visit of Russian President Vladimir Putin to South Korea in November 2013. Several options for electricity transmission from Russia to NEA countries were considered:

- Transmission line from Vladivostok transiting through the DPRK to Seoul;
- Transmission line from Chita and Blagoveshchensk transit through China (Harbin, Shenyang) and DPRK to Seoul;
- Transmission line from Chita and Blagoveshchensk transit through China (Harbin, Shenyang, Dalian) and then through an undersea cable to Seoul.

It was noted that Japan's participation would enhance the economics of the project. Electricity can be transmitted via the northern route (submarine cable from Sakhalin), and the southern route (from the Republic of Korea through a submarine cable via the island of Kyushu to Honshu).

"Gobitec" and the Asian Super Grid. The "Gobitec" is a new industrial initiative, in which electricity produced from renewable energy sources in the desert regions of Mongolia and China is brought to the industrial centers of Mongolia, China, the Republic of Korea, and Japan via high-voltage lines—the Asian Super Grid

(ASG).[19] The Asian Super Grid Initiative aims to construct high-voltage transmission lines throughout the NEA region and interconnect the national grids of Japan, the Koreas, China, Mongolia, and Russia so that abundant renewable energy sources in remote areas can be utilized, such as hydropower resources in East Siberia.

The proposed grid connects renewable energy sources in the Gobi Desert with Irkutsk to the north, incorporating hydropower electricity into the system. It connects these sources to the locations of demand to the East, including Shanghai, Seoul, and Tokyo.

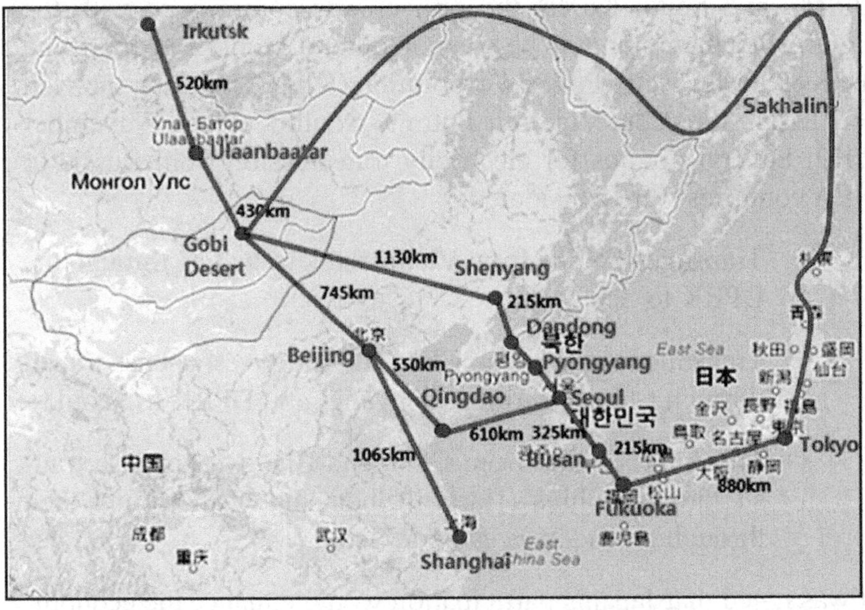

Figure 12. Possible Routes of the ASG.

In March 2013, five partner organizations signed a memorandum of understanding to jointly prepare a regional study on Gobitec and the Asian Super Grid. They include the Energy Charter Secretariat, the Energy Economics Institute of the Republic of Korea, the Ministry

19. The Gobi Desert is estimated to be the third largest potential source of solar energy in the world and is also blessed with steady, strong wind speeds, making it ideal for both technologies.

of Energy of Mongolia, the Japan Renewable Energy Foundation, and the Energy Systems Institute of the Russian Federation.

The core objective of the joint study was to provide NEA countries and the international community with reliable information and data concerning the potential of renewable energy sources in the region. The study also aims to stimulate the interest of the private sector and international investors, and to provide recommendations for maximizing international and regional cooperation in order to promote the Gobitec and the Asian Super Grid initiative. The study was completed in 2013 and presented to interested policy makers, the business sector and the general public to stimulate informed discussions and facilitate regional cooperation in the energy sector in NEA.

The super grid entails technical and political challenges because of its large geographical size. One advantage, however, is that demand and supply can be balanced more easily because regional differences are leveled out by the size of the system. Also, energy conversion technologies can be concentrated in places of high energy output, while the grid provides connection to the places with high electricity demand. The ASG will allow for the free exchange of electricity between countries in NEA, thus providing large economic, social, and environmental benefits.

Energy Cooperation under Multilateral Mechanisms in NEA

Recognizing the necessity for creating an institutional mechanism for multilateral cooperation in the region, the countries in NEA began to move towards the establishment of regional cooperative bodies in the mid-1990s. Those efforts include the Great Tumen Initiative and the Intergovernmental Collaborative Mechanism on Energy Cooperation in North-East Asia.

Greater Tumen Initiative. The Greater Tumen Initiative (GTI) is an intergovernmental cooperation mechanism among four countries: China, Mongolia, the Republic of Korea, and the Russian Federation, supported by the United Nations Development Program (UNDP). In 1995, GTI member governments signed agreements establishing this mechanism with the aims of strengthening economic and technical cooperation and attaining greater growth and sustainable development in NEA and especially the Greater Tumen Region (GTR). The GTR includes China's three northeast provinces and Inner Mongolia, the three eastern provinces of Mongolia, the eastern ports of the ROK, and the Primorsky Territory of Russia.

Energy is one of the GTI's priority areas for regional cooperation. The GTI Energy Board was established in 2009 to address regional energy issues on a policy and strategy basis while also including private sector expertise and resources. The following objectives for energy sector cooperation have been identified:

- Enhancement of energy policy coordination and cooperation;
- Reduction of non-physical barriers for energy trade and investment in the GTR; and
- Promote exchange of information on energy among GTI member countries.

Figure 13. The Greater Tumen Region (GTR).

Intergovernmental Collaborative Mechanism on Energy Cooperation in North-East Asia. South Korea has taken an active initiative to create an intergovernmental cooperation mechanism among the countries in NEA since 2001.[20] At the June 2001 International Symposium on Energy Cooperation in NEA, the ROK Ministry of Commerce, Industry and Energy (MOCIE) proposed the creation of a senior officials committee to elevate current energy cooperation issues to the intergovernmental level. As the first step toward the creation of this collaborative framework, an intergovernmental meeting was held in Khabarovsk, Russia in October 2001 and was attended by government officials and experts from the six countries in the region, namely, China, Japan, Mongolia, Russia, the DPRK and the Republic of Korea. At the meeting, the Khabarovsk Communiqué, which contains the objectives and basic principles of energy cooperation in Northeast Asia, was adopted.

20. Keun Wook Paik, "Towards a Northeast Asian Energy Charter," *Energy Policy* 20, No. 5, May 1992, pp. 433-443.

The agreed objectives for energy cooperation are to increase the supply of energy from the NEA region; to optimize the efficiency of supply and use of energy; and to minimize the environmental impact of energy projects through improved energy mix. The basic principles manifested in the Communiqué include: The recognition of sovereign rights over energy resources; development of free and fair trade; promotion and protection of investment and environmental protection; and the free and non-discriminatory transit of energy products. The Khabarovsk Communiqué also recommended the creation of a "Senior Officials Meeting" among the six countries, a secretariat, and five 'working groups' on areas such as energy planning, programming and restructuring, electric power interconnection, and the interstate transit of fossil fuels.[21]

After several subsequent preparatory meetings, the first Senior Officials Committee (SOC) Meeting, held on November 16-17 in Ulaanbaatar, Mongolia, adopted the Ulaanbaatar Statement of Senior Officials on Energy Cooperation in North-East Asia for the creation of the Intergovernmental Collaborative Mechanism on Energy Cooperation in North-East Asia with the DPRK, Mongolia, the Republic of Korea and the Russian Federation as members. Under this mechanism, the Working Group on Energy Planning and Policy (WG-EPP) was created in order to implement cooperative activities among the member countries. China also joined the WG-EPP activities, although as an observer of the SOC rather than a member. As part of the activities for regional cooperation, the WG-EPP prepared and published "Energy Policy Survey" and "Energy Statistics Data Base" country reports for China, Mongolia, the Republic of Korea and the Russian Federation. In 2007, the group developed a "Regional Energy Demand/Supply Outlook for Northeast Asia" report and organized the first Government–Business Dialogues (GBD). In 2008, the WG-EPP undertook a joint survey study, "Energy Production Potential and Development Plans in the Northeast Asian Countries."

---

21. Keun Wook Paik, "North Korea and Seabed Petroleum," Washington, DC: Woodrow Wilson International Center for Scholars, 2007, available from *www.wilsoncenter.org/publication/north-korea-and-seabed-petroleum*.

Strategy to Facilitate Regional/Multilateral Energy Cooperation in NEA

Creating Consensus. The starting point for energy cooperation in NEA can be derived from a common recognition that the region is a single community in terms of its energy use and environment, and that its energy and environmental resources are common assets, which are essential for facilitating economic growth and prosperity, not only for our generation but also for future generations. Establishment of a basis for mutual cooperation cannot be successful without sharing mutual interests and exploring joint endeavors to pursue common prosperity. In initiating regional energy cooperation, the countries in the NEA region need at least to reach a consensus for common goals of:

- Removing or reducing impediments to the efficient development, production, supply and use of energy;
- Minimizing the adverse environmental consequence of energy development, production and use; and
- Enhancing regional energy security with common task-sharing efforts.

Lessening Political Uncertainty. Political stability should be pursued in a multilateral cooperative framework, as it is an important factor in energy projects, particularly, related to the security of the projects. Uncertainty with political stability involves a high risk for investors in the energy industry. In order to facilitate the energy cooperation in NEA, the countries in the region need at least to reach a consensus for common goals of lessening political risk of the energy projects and de-coupling energy issues from the politics.

Identification and Prioritization of Cooperation Projects for Common Interests. This includes multilateral development of energy resources, oil, natural gas, and coal, as well as the construction of energy network systems for cross-border pipelines and power grids for both short and long-term basis and for the multilateral level for common interests among the countries in NEA. Also, energy technology cooperation for the improvement of energy efficiency or energy conservation, and the development of renewable energy should be of high priority for regional cooperation among the countries in the NEA region. Thus the priority areas include:

- Joint efforts for development of oil, gas, and coal resources in NEA
- Utilization of renewable energy and improvements in energy efficiency
- Oil and gas pipelines and grid interconnection
- Development of policy, regulation for energy trades within the region

Enhancement of Policy Cooperation among the Government and Business Sectors. To enhance policy cooperation for the energy sector among the countries in NEA, the following cooperation agendas need to be defined for the implementation stage:

- Establishment of a mechanism for energy data and market/industry information sharing
- Capacity building to enhance the market and policy compatibility between the countries
- Encouraging private sector and industry participation
- Decoupling energy projects from the politics to ensure market transparency for domestic/foreign investment and also affordable pricing mechanism for the long-term contracts

Since each country in the region has its own energy data or statistics system, which is not mutually compatible with other countries' systems, it is necessary to develop a common database for the energy development projects in each country, for both the upstream and downstream sectors of oil, gas, coal, and electricity industries. Priority areas include:

- Developing and regularly updating common energy databases and policy survey/ sharing dialogues at the multilateral level
- Expanding the number of participating countries by adding Japan and the DPRK
- Establishing and maintaining a communication mechanism

Capacity building is required for regional energy cooperation in NEA because the countries in the region differ from each other in terms of economic systems and the level of economic development. Therefore maintain different market systems including in the energy sector, so that the countries in the region need to make an effort to enhance market compatibility in order to implement energy cooperation at the regional level. Priority areas include:

- Facilitation of capacity building programs
- A broad network among policymakers, industry, researchers and investors
- A policy dialogue forum (GBD to be enlarged to encourage the business participations)

Improving the Investment Environment. Exploration and development of oil, natural gas, and coal and the construction of the necessary transportation infrastructure requires a massive investment. However, the countries in the region with abundant energy reserves, like Russia and Mongolia, suffer from tight capital markets for financing these energy projects. Therefore, the use of multilateral cooperative approaches to create a more favorable environment for foreign energy investment in the region will be necessary. Priority areas include:

- Creating a more favorable environment and market-friendly policies for attracting foreign investment in Russia and Mongolia

- Actively encouraging public-private partnerships (PPPs) for investment in energy development
- Promoting joint ventures with foreign companies and use of funds from multilateral development banks or international financial institutions to facilitate investment in large-scale energy projects
- Establishing inter-governmental committees for supporting energy infrastructure that conduct financial reports, technological inspections, and feasibility studies in a consistent way
- Establishment of a regional multilateral development bank such as the Northeast Asian Development Bank and the Asian Infrastructure Investment Bank for the countries in the region[22]

Roadmaps Towards Multilateral Energy Cooperation in Northeast Asia

Based on the policy agenda identified in the previous section as well as on the survey on experiences of successful regional energy cooperation in other regions, the policy agenda to facilitate regional energy cooperation in the NEA region can be implemented in two stages for a gradual step-by-step approach; one is to be pursued in the short term, and the other is for the long term.

22. In March 2014, ROK President Park Geun-hye during her visit to Germany announced that the government would provide support to the DPRK by establishing an NEA Development Bank if the North decides to give up its nuclear weapons. The idea was first introduced in the NEA Economic Forum Meeting in Tianjin on September 2-7, 1991, by the former ROK Prime Minister Duck Woo Nam. In October, 2013, Chinese President Xi Jinping suggested the establishment of Asian Infrastructure Investment Bank during his tour to Southeast Asian countries, and the Chinese government has discussed its establishment with many countries, including the countries in the NEA region.

The short term agenda, which can be implemented quickly, is a soft policy agenda on non-binding manners, including:

- Establishment of a channel for policy dialogues between governments in the NEA region to foster confidence building
- Promotion of information and data exchange and sharing mechanisms
- Joint research/study to identify possible cooperative energy projects (i.e. natural gas pipelines, power interconnection, oil stockpiling)
- Capacity building projects for the less-developed countries in the region
- Encourage energy expert/business dialogues & participation
- Assistance to and cooperation with the DPRK to resolve its energy shortage problems[23]

The long-term agenda, which would require a consensus as well as more preparatory joint efforts between the countries in NEA, include:

- Creation of institutionalized frameworks for multilateral energy cooperation by enacting a treaty, charter, or regional energy community at the regional level
- Introduction of policy coordination functions within the established institutional arrangement
- Development of joint policy agenda for common goals/task sharing
- Addressing intra-regional energy financing mechanisms

23. Selig S. Harrison, Ji-Chul Ryu, and Enoji Kyongje Yon'guwon, "Energy Aid for North Korea: Fossil Fuel Options and Denuclearization," Ulsan, Republic of Korea, Center for Energy Research, North Asia, Korea Energy Economics Institute, 2007.

Conclusion

This chapter surveyed and discussed the current status and future of energy in each country—South Korea, China, and Japan—in an effort to derive a number of policy implications and recommendations for each nation to pursue towards the establishment of a sustainable energy system. Then, it identified various issues related to energy cooperation in NEA and also developed a policy agenda to facilitate regional energy cooperation in the region.

Given the political uncertainty in NEA, regional energy cooperation may be a difficult challenge, at least in the short term. However, energy cooperation in NEA is an urgent need in order to stabilize the regional energy supply and demand structure. However, energy development in NEA may need a high investment cost. Given the advantage of geographical proximity of large energy producing areas to large consuming areas and its symbolic nature to enhance the political stability and economic prosperity of the region, we should start to devote our efforts to develop a mechanism to facilitate energy cooperation between the countries in the region in pursuit of a common benefit for all the regional countries.

The problems that the world energy market is facing today can neither be confined to a single country nor to a single region; they are those of all nations. No country can remove itself from the problems involved with energy development and environmental degradation. Energy resources and environment conservation have proven to be one of the essential means to promote a task sharing for the regional energy security of countries in Northeast Asia.

Chapter 3

Japan's Electricity System: Reform After Fukushima

Hiroshi Takahashi

Introduction

Following the Fukushima Daiichi nuclear power plant incident of March 2011 (hereafter referred to as "Fukushima"), Japan's energy policy is in need of a radical overhaul. The first issue is the mid- to long-term discussion regarding Japan's energy mix, specifically what to do with the country's nuclear power plants. In September 2012, the Democratic Party of Japan put forth the "Innovative Energy and Environment Strategy," which included abandoning nuclear power, but with the change of administrations at the end of December of the same year that strategy was scrapped. Subsequently, in April of 2014, the Cabinet of Prime Minister Shinzo Abe approved the "Basic Energy Plan 2014," thereby reopening the way towards a return to nuclear power, but the contents of that plan are broad and give little in the way of direction.

The second issue, and the main topic of this paper, is the reform of the electricity system. This would involve unbundling transmission grids from vertically integrated utilities and completely liberalizing the retail market in order to allow the large-scale deployment of renewable energy sources (RES) and give consumers more choices. This endeavor has continued even after the change of administration, with the Abe Cabinet steadily pushing forward with amending the Electricity Business Act. Reforming the regulations surrounding the current monopolistic electricity industry will provide new business opportunities, and many companies are preparing to enter the market.

When it comes to electricity reform, the key phrase is distributed energy sources (DES). In the past, most electricity systems around the world have been comprised of large-scale power companies, which combine power generation and transmission, developing nuclear and large-scale coal power plants (i.e., centralized power generators), and long-distance transmission grids built systematically from those locations. In contrast to such centralized systems, a new system is appearing around the world—a distributed electricity system in which various new market entrants develop RES, companies install co-generators in order to supply their own energy needs, and demand response (DR) aggregators stimulate so-called "negawatts" from customers.[1]

For a long time it was believed around the world that centralized electricity systems were superior, but from the 1990s on, distributed systems began to be revisited by Western countries. The reason: A changing environment due to electricity market liberalization and a call for lower carbon emissions. Japan's electricity system reform is an attempt to reassess the overly centralized system in response to the historic incident at Fukushima. A shift towards a distributed system would make it difficult to maintain or develop nuclear power, affecting Japan's electricity structure and energy security as a result. And if such reforms become the global norm, then Japan's efforts will no doubt have a ripple effect on the rest of Asia as well.

The purpose of this paper is first to understand Japan's electricity system reform before considering the effects on a distributed elec-

---

1. "Negawatts" is a notion advocated by Amory Lovins. If consumers could save electricity effectively in response to supply shortage, the "minus supply" would be an equivalent to costly additional power generation. See Amory B. Lovins, "The Negawatt Revolution," *Across the Board* XXVII, no. 9, September 1990, available from *https://www.rmi.org/wp-content/uploads/2017/06/RMI_Negawatt_Revolution_1990.pdf.*

tricity system, and finally providing an outlook of the future. The first section examines Japan's electricity crisis following Fukushima and lays out the details of electricity system reform. The second section first defines DES before identifying the need to shift to such an electricity system and concomitant challenges. The third section introduces the current state of DES in Japan and its future potential, after which it examines the effects of Japan's energy policy, including electricity system reform. The final section gives an outlook on DES in East Asia.

Electricity Crisis After Fukushima and Electricity System Reform in Japan

Centralized Electricity System and Electricity Liberalization

Before 1990, most countries had electricity systems with legal monopolies and vertical integration from power generation to transmission and retail. Because the economy of scale works in such a system, the power company with monopoly was able to invest heavily in centralized power sources like nuclear and large-scale hydroelectric plants and long-distance transmission grids, and recoup its investment through regulated prices. As a result, self-generated power was uncommon, other than for in-house or emergency use, and the introduction of renewable energies was severely limited. In a centralized electricity system, it was very difficult and uneconomical for anyone but the monopoly-holder to own power plants and sell the electricity generated there.

From the 1990s onward, however, electricity liberalization became a strong trend among developed nations. While small gas turbines became more economical, it also became easier to integrate and control diverse power generators as information technology (IT) advanced. Consequently, opening up the generation and retail markets and having consumers behave based on price indices allowed the resources of the whole system to be distributed more efficiently. These were the underlying factors of electricity liberalization.

In Great Britain, as the administration of Prime Minister Margaret Thatcher pushed through new liberal reforms in rapid succession, the Central Electricity Generating Board had its transmission separated from other businesses,[2] and was taken private in 1990, and the generation and retail markets were liberalized. Norway and Sweden took similar methods in 1992 and 1996, respectively. The United States' electricity system varies by state, but such large states as New York, Texas, and Pennsylvania are making progress in liberalization. As the economy of self-generation increased, owners of such generators sought entry into the generation market, and retail was also liberalized not long thereafter.

In contrast, Japan has been the wariest of all developed countries when it comes to electricity liberalization. The system of ten regional companies with vertical integration and legal monopolies, formed following World War Two (WWII), was able to cope with the ravenous demand for power during Japan's period of high economic growth. After 1995, liberalization took several steps forward, but the incumbent power companies' opposition was strong and competition policy was not exercised thoroughly. The number one reason for the power companies' opposition to liberalization was stable power supply. They claimed that grid unbundling for the sake of competition would make it difficult to coordinate between the two and would lead to power outages. Their second reason was nuclear power. They asserted that liberalization would lead to the prioritizing of short-term profits, and that power sources which benefitted

---

2. There are three forms of grid unbundling. Ownership unbundling, generally used in Europe, involves cutting off the transmission section of the integrated power company in terms of capital. Functional unbundling, commonly used in the United States, involves surrendering the operations rights to an independent system operations body, regardless of who owns the transmission grid. Legal unbundling, used in France and some parts of Germany, involves designating the power company as a holding company and the transmission section as its subsidiary. The strongest structural measure is ownership unbundling.

the public, such as nuclear power, would become unsustainable.[3]

In 2003, when the attempt at unbundling failed in the so-called "third-stage liberalization" as outlined above, electricity liberalization faded from the forum of serious discussion in Japan. As a result, almost 40% of all demand was still filled by legal monopolies in the small retail market, and even in the liberalized bulk market, which filled more than 60% of demand, the share of new entrants was no more than 3.5% in 2011 according to the Agency for Natural Resources and Energy's (ANRE) Electricity Survey Statistics.[4] Japan's effective monopolies from before Fukushima had continued until April 2016, when the entire retail market was liberalized.

Fukushima Daiichi Nuclear Incident and the Electricity Crisis

The Great East Japan Earthquake of March 11, 2011, not only caused a hydrogen explosion at the Fukushima Daiichi nuclear power plant and the resulting radiation contamination of the environs, it also brought about an electricity crisis, in which supply fell to critical levels. The magnitude 9.0 earthquake and 10-plus meter-high tsunami rendered utterly inoperable not only Fukushima Daiichi (six reactors totaling 4.7 gigawatts [GW]), but also other centralized installations arrayed along Japan's Pacific coast, including the Fukushima Daini nuclear power plant (four reactors totaling 4.4GW), the Hirono thermal power plant (five generators totaling 3.8GW), and the Kashima thermal power plant (six generators to-

3. In the Agency for Natural Resources and Energy's (ANRE) Electricity Business Subcommittee Meeting report which discusses unbundling, it states: "In order to advance large-scale power generation businesses such as nuclear power, they must be combined with transmission businesses," and that "even if retail liberalization progresses further, we must create an environment in which nuclear power can be pushed forward." See, Agency for Natural Resources and Energy, "On the Future Desirable Framework of the Electricity Business," February 18, 2003.

4. Agency for Natural Resources and Energy, "Electricity Survey Statistics," March 31, 2014.

taling 4.4GW). The Tokyo Electric Power Company (TEPCO), Japan's largest electrical company, had more than 60GW of installed capacity, but three days after Fukushima, on Monday, March 14, it could muster no more than 31GW.

TEPCO called for the aid of its counterparts across Japan, but the transmission grid connecting TEPCO's region with that of Chubu Electric Power has a capacity of only 1GW—a drop in the bucket. Although western Japan had not been severely affected by the disaster and had tens of gigawatts of electricity to spare, eastern Japan could not make use of the power.

Consequently, TEPCO enacted rolling blackouts on March 14, 2011, for the first time since WWII. By subdividing its supply regions and stopping electricity to each area in a set order and by time period, TEPCO avoided large-scale, unpredictable power outages. However, because there was little time to prepare for and publicize these rolling blackouts, many people were irked by the unilateral and across-the-board measures taken by TEPCO.[5] Much criticism suggested that if pricing plan menus, which encouraged peak shifting, had been widespread, and if smart meters had been installed in homes, then energy would have been saved more effectively based on price mechanisms, and the effects of the rolling blackouts would have been ameliorated.

As thermal power plants were brought back online and the cold winter temperatures started to rise, power consumption dropped and the rolling blackouts ended on March 28, 2011. However, with the annual record-high demand peaks of Japan's hot summer fast approaching, there were still concerns about supply shortages within TEPCO's region. To alleviate this concern, the government enacted an Electricity Usage Limitation Ordinance based on the Electricity Business Act, which required large consumers to cut peak demand by 15% of the previous year's. As a result, during the summer of 2011, peak demand in TEPCO's region fell 18% from 59.99GW

5. On March 17, 1.8 million households suffered rolling blackouts.

the year before to 49.22GW, with a large reduction of 20.7% in the daily average peak demand (between July 1 and September 9) from 51.44GW the year before to 40.8GW.

With such enormous peak cuts, TEPCO was able to avoid both unexpected and planned power outages. However, using an Electricity Usage Limitation Ordinance, which carries penalties for transgressions, had severe effects on the production plans of factories, and criticism of TEPCO increased. With across-the-board peak cuts of 15%, TEPCO had again failed to use a method that allowed flexibility in energy savings between companies which found it easy and those which found it difficult to abide by the ordinance.

Move Towards Electricity System Reform after Fukushima

Policy-makers and power companies were given a shock by many factors related to the fallout from Fukushima: Over-reliance on centralized power generation had caused a supply shortage following the disaster; the transmission system connecting the entire country had not functioned effectively, and they had been unable to adjust supply to meet demand flexibly through market mechanisms. This was because the very same region-based monopoly system, which they had protected under the argument of stable supply being paramount, had inhibited the all-important stable supply of electricity. With the nuclear power plant incident and the criticism of the power companies by the Japanese people weighing on them, the ruling Democratic Party of Japan set about reforming the electricity system.

In February 2012, the Agency for Natural Resources and Energy established the Electricity System Reform Expert Subcommittee and began discussing reform proposals. Normally, executives of several power companies would be present, but this time they were excluded; the majority of committee members were economists and other reformists. The midterm report, delivered in July 2012, included an idea from 10 years earlier: Grid unbundling. Thereafter, the committee continued discussions, and following the change of administrations in December, it delivered the final report to the Abe administration in February 2013.

The goals of the reform outlined at the beginning of the report were fourfold: (1) To expand introduction of renewable energies, which had been severely lacking in the past; (2) to promote cross-regional supply-demand adjustment; (3) to give households and other consumers choice in power companies and services; and (4) to make use of demand response. To summarize, the goal was to account for variable power sources and widely changing power structures while building a stable power supply system that makes use of market mechanisms.

The committee's detailed plan can be summarized in three points. First is the establishment of the Organization for Nationwide Coordination of Transmission Operators. After reflecting on the electricity crisis, the committee decided to establish a national organization (incorporated body) which would coordinate system operations and grid development plans among power companies across the country, thereby promoting a flexible electricity supply between regions. In 2013, the Electricity Business Act was amended, with the establishment of this organization planned for 2015. Once this is accomplished, large wind-farms can be built in Hokkaido, which has an excellent wind environment, and the resulting power can be sent to Japan's main island of Honshu, where supply-demand coordination can be carried out with ease.

Point number two is the complete liberalization of the retail market. The June 2014 amendment to the Electricity Business Act has completely opened all retail markets, including the households market that had been under monopoly until April 2016, providing consumers with the ability to choose. Not only will the new competition drive down prices, but a greater variety in price plans will allow expanded smart demand response, and new services in combination with electric vehicles and fuel cells will become available.

Finally, the committee set forth grid unbundling as a structural competition policy. The proposal is to make the original vertically integrated power company into a holding company and name the transmission section as a subsidiary or separate company. This legal unbundling would take effect no later than 2020 according to the amendment to the Electricity Business Act. This would effectively render the transmission company neutral and remove the issues of renewable energies and new market entrants connecting their generation facilities to the power grid. At the same time, a real-time balancing market would be created and supply-demand would be coordinated efficiently based on market principles.

These reforms would not simply stop at liberalization, but may open the way to the distribution of the entire system. Here let us switch focus and look at the DES of developed countries and how Japan's electricity system reform relates to them.

Distributed Energy Sources and Electricity System Reform

What Are Distributed Energy Sources?[6]

There are many possible definitions of DES or distributed generation.[7] European Union (EU) Directive 2003/54/EC concerning

6. For previous comprehensive research on DES, see, Amory B. Lovins, *Small is Profitable*, Snowmass, CO: Rocky Mountain Institute, 2002; Jeremi Martin, *Distributed vs. centralized electricity generation: are we witnessing a change of paradigm?* Paris: HEC Paris, 2009; and John P. Banks, et al., *Assessing the Role of Distributed Power Systems in the U.S. Power Sector,* Washington, DC: Brookings Institution Energy Security Initiative and Stanford, CA: the Hoover Institution Shultz-Stephenson Task Force on Energy Policy, 2011.

7. For previous discussion on definition of DES, see, Thomas Ackermann, Goran Andersson, and Lennart Soder, "Distributed Generation: a definition," *Electric Power Systems Research* 57, no. 3, April 2001, pp. 195-204; and Guido Pepermans, Johan Driesen, Dries Haeseldonckx, Ronnie Belmans, and William D'haeseleer, "Distributed generation: definition, benefits and issues," *Energy*

common rules for the internal market in electricity defines them as "generation plants connected to the distribution system."[8] On the other hand, a 1999 report from the International Conference on Electricity Distribution (known by its French acronym CIRED) focuses on generation capacity up to 50MW or 100MW at most.[9]

In this paper, we will use the following four categories. First, historically speaking, the term "DES" has referred to electricity self-generation. In contrast to the large-scale generators of power companies, businesses will sometimes install small or mid-sized thermal generators in their factories to supply their own inexpensive power. Called "customer-owned power generators" by big power companies, these setups experience almost no power loss due to transmission or distribution because the generator is on-site; they allow for cogeneration and they are excellent for facilitating business continuity in emergency situations. The falling prices of small gas turbines and innovations in cogeneration technology spurred private generators usage and from the 1990s onward they have gained in popularity.

The second category are renewable energy sources (RES). The falling prices of wind turbines and solar panels from the 1990s resulted in RES becoming a global trend (Figure 1). Global warming and the introduction of policies prioritizing and subsidizing renewable energies in many countries also contributed to this trend. In addition to highly variable power sources such as wind and solar, biomass, geothermal, solar thermal and small hydroelectric energies currently in use, there is hope that we will be able to use wave and tidal energies in the future with the necessary technological advances.

Policy 33, no. 6, April 2005, pp. 787-798.

8. European Union, "Directive 2003/54/EC of the European Parliament and of the Council," *Official Journal of the European Union*, L 176/42, Luxembourg: Publications Office of the European Union, June 26, 2003, available from *https://eur-lex.europa.eu/legal-content/ENG/TXT/PDF/?uri=OJ:L:2013:176:FULL.*

9. N. Jenkins, et al., "CIRED Working Group 4 on Dispersed Generation," preliminary report prepared for CIRED 1999, the 15th Annual International Conference and Exhibition on Electricity Distribution, Nice, France, June 1-4, 1999.

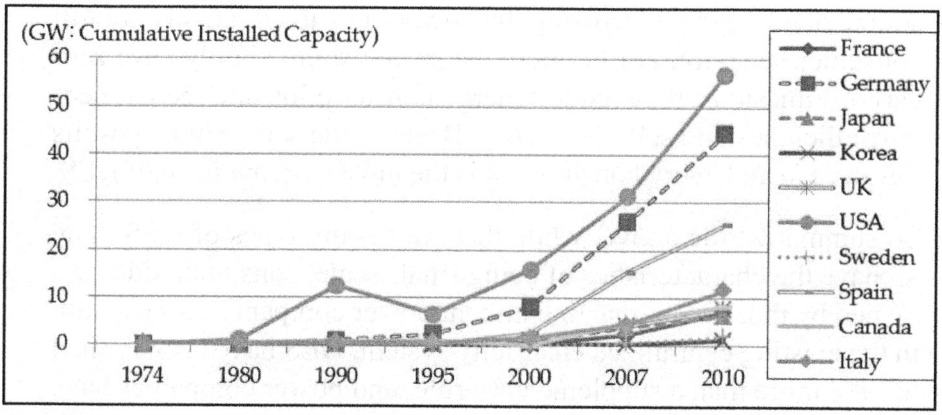

Figure 1. Renewable Energy (Except Hydroelectric) Installed Capacity of Leading Countries.[10]

The third possible category is fuel cells and storage batteries, a recent technological innovation expected to spread in the future. Fuel cells produce electricity by taking natural gas, reforming it into hydrogen gas and causing a chemical reaction. The advantages of fuel cells are that they can be used in cogeneration, their energy efficiency can be increased, and they produce relatively little carbon dioxide (CO_2). Furthermore, batteries that store energy but do not generate electricity can also be considered as part of DES. In the past, storing electric energy was very difficult, but if large capacity lithium-ion batteries and sodium-sulfur (NaS) batteries continue to become less expensive, they will become ubiquitous in the electricity system and play a large role in supply-demand adjustment.

The fourth and final possible interpretation of DES is demand-side coordination, or demand response (DR). For example, if consumers cut peak usage by a certain amount when supply-demand is tight, that is of equivalent value to the expensive extra power generation that would have been required to meet that demand. This concept, called "negawatts" by Amory Lovins, could provide a valuable

---

10. International Energy Agency, *Electricity Information*, 2012 edition, Paris: Organization for Economic Cooperation and Development/International Energy Agency, 2012, p. III.8.

source of energy by borrowing the power of IT to effectively control consumer behavior. Furthermore, negawatts don't simply have zero carbon emissions, they reduce energy consumption and increase energy efficiency and self-sufficiency. Because the source of negawatts lies in each and every consumer, it is the most extreme form of DES.

To summarize the above, while there are many types of DES, they all have the characteristics of being small-scale, consumer-side, and owned by those other than incumbent power companies. As a result, in the existing centralized electricity system, DES have been limited to little more than a supplementary role, and power companies have not actively pursued their development. However, recent changes in the power generation environment, such as cost reduction through technological innovation, increased economic value due to liberalization, and the call for lower carbon emissions, have caused many to revisit DES as a possibility for the future.

Distributed Electricity System for Distributed Energy Sources

The biggest reason that monopolistic power companies have not pursued DES is that it threatens to alter their existing business model and would require a new distributed electricity system.

First, grid unbundling would be essential. During the period of centralized energy dominance, only monopolistic power companies were able to develop large-scale power generation plants and there was no need to allow others to use the transmission grid. However, if various new entrants were to sell their self-generated power, or if RES were introduced into the system, the transmission grid would have to be made open to all. In Western countries, the structural measure generally used to achieve this is unbundling, but the vertically integrated power companies are obviously opposed to this.

Second, the transmission and distribution grid would need to be expanded and upgraded. A centralized energy system has a relatively limited number of large-scale power sources, and as such monopolistic power companies are able to plan out the transmission grid with one eye on the locations of those plants. In other words, the existing grid is optimized for the existing centralized electricity system. However, a distributed electricity system is spread out geographically, and so the transmission grid would have to be expanded into areas it was not previously needed, such as northern Hokkaido. Additionally, with the recent advances in IT, grid operators have become able to directly control dispersed wind generator outputs and DR has become very effective through the use of smart meters. The question of who will take the responsibility and burden for investing in such technologies will become a significant issue.

Third, how the grid is operated will have to change. For example, Germany's renewable energy law stipulates that renewables are prioritized for connection and dispatching. Also, with the increase in the number of variable power sources, the use of adjusting power supplies will change, too. Not only will grid operators face new challenges to provide a stable power supply, but power generation businesses will face the issues of falling use rates for gas thermal power and the need for output-adjusted operation of nuclear and coal thermal plants as well.

Finally, the design of the market would change. Grid unbundling would mean that grid operators would no longer have generation facilities, so any supply-demand adjustment would in principle be done through the market. Supply and demand would be roughly matched based on the previous day's spot market, with final adjustments being made on today's real-time balancing market. Such markets would need to be created and the setups of both grid operators and power generation businesses would change. New players, such as traders and DR aggregators, would need to be involved in these markets. There is also discussion of the need for a capacity mechanism for thermal generation plants, which are seeing reduced capacity factors.

As we have seen, the expansion of DES calls for the distribution of the electricity system and calls for a radical shift in existing business models. In Germany, the government's policy shifts away from nuclear power and towards renewables and electricity liberalization have led not only to the corporate breakup of incumbent utilities, but to their falling behind in investing in renewables and the falling capacity factor of gas thermal generation.[11] In order to give DES a policy advantage, large and systematic policy change is necessary.

Issues with Distributed Energy Sources

DES are being re-evaluated in more and more leading developed countries, with European countries in particular, such as Germany, Denmark, and Spain, steadily making progress by giving it policy priority. Whether from the standpoint of energy self-sufficiency, climate change, safety against terrible disasters, next generation technologies, or international competition, there is hope that distributed will replace centralized energy systems in the future. However, there are many uncertainties and obstacles along that path.

The first obstacle is economic. DES are still less economical than centralized power sources because they do not follow the rule of economies of scale. Among RES, wind power has fallen in price significantly, but fuel cells still lack competitiveness. While DR is being applied practically in the United States, a structure for effectively gathering negawatts has yet to be established. Until we see further technological innovation and volume efficiency, it will be economically unfeasible to depend on DES.

Second is the technological uncertainty of system reform. As mentioned above, in order to reform towards a mainly distributed electricity system, one must first provide effective measures to stabilize variable power sources and come up with a DR program using market mechanisms. Variable power sources account for approximately 30% or more of Germany's and Denmark's power mix and

---

11. E.g., "How to lose half a trillion euros," *The Economist*, October 12, 2013, available from *https://www.economist.com/briefing/2013/10/15/how-to-lose-half-a-trillion-euros.*

they have yet to experience any significant operation problems, but there is no guarantee that this will hold true at 80%. Furthermore, expanding the transmission grid and establishing hydrogen infrastructure for a distributed system would require enormous investment. Who would take such risks, and how? What would the timeline of reforms look like? These issues are still being discussed in many countries.

Third is the issue of what to do with existing power companies after such a momentous structural shift. In other words, how can the opposition of the incumbent power companies be restrained and how can they be made to step down peacefully? The reason Japan was not able to achieve unbundling was because the opposition of the power companies was just too strong. Additionally, giving favor to distributed power sources means that centralized power will become disadvantaged. After electricity began to be liberalized in the 1990s, many developed countries saw a plateau in the number of new nuclear power plants being built (Figure 2). In response to this, the British government introduced "Contracts for Difference" for new nuclear plants. It will be difficult to develop DES without finding such a balance of policies.[12]

---

12. "Contracts for Difference" (CfD) is a policy instrument to support low-carbon power sources by guaranteeing the selling price of such electricity for 15 years for RES and for 35 years for nuclear. See Department of Energy and Climate Change, "Electricity Market Reform: Delivering UK Investment," Cm 8674, London: The Stationary Office Ltd., June 2013, available from *www.gov.uk/government/uploads/system/uploads/attachment_data/file/209276/EMR_Spending_Review_Announcement_-_FINAL_PDF.pdf*.

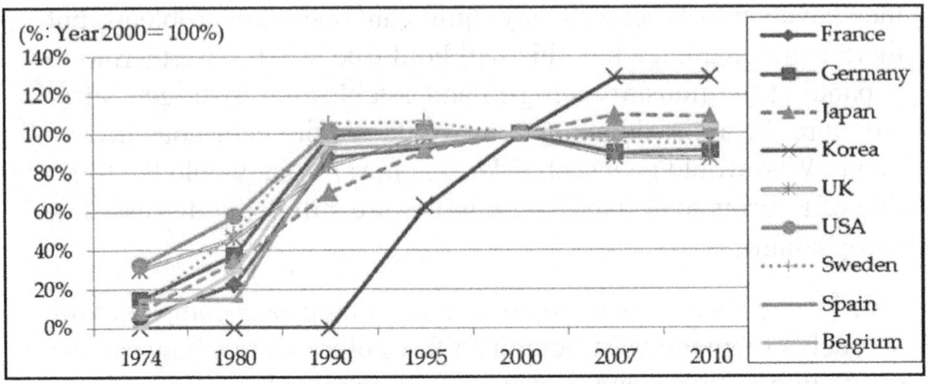

Figure 2. Installed Capacity of Nuclear Power Plants in Leading Countries.[13]

Potential of Distributed Energy Sources in Japan

State of Distributed Energy Sources in Japan and Related Industrial Infrastructure

There is no clear agreement in Japan on whether the country can make the shift to DES and a distributed electricity system as outlined above. Japan has not introduced very much renewable energy (Figure 1) and electricity liberalization has made no headway. Nevertheless, the electricity system reform that the Japanese government is currently pursuing seems to be aiming towards a distributed system. If so, does Japan, in fact, have the necessary technological and industrial wherewithal to actually make use of DES?

First of all, Japan boasts the most electricity self-generation in the world, with 3,963 facilities producing 57.70 GW (Figure 3). This number is greater even than the installed capacity of domestic nuclear plants and has been growing steadily over the past several years. While many of the generators are thermal ones, this number is likely

13. International Energy Agency, *Electricity Information*, 2012 edition.

to continue growing as the memory of the rolling blackouts after Fukushima is still fresh in people's minds. Approximately half of self-generation capacity is consumed in-house, while the remaining half is sold off. Self-generated electricity has been an important power source for new power companies to date, and as liberalization makes inroads, the increase in self-generation will help with the growth of the distributed system as well.

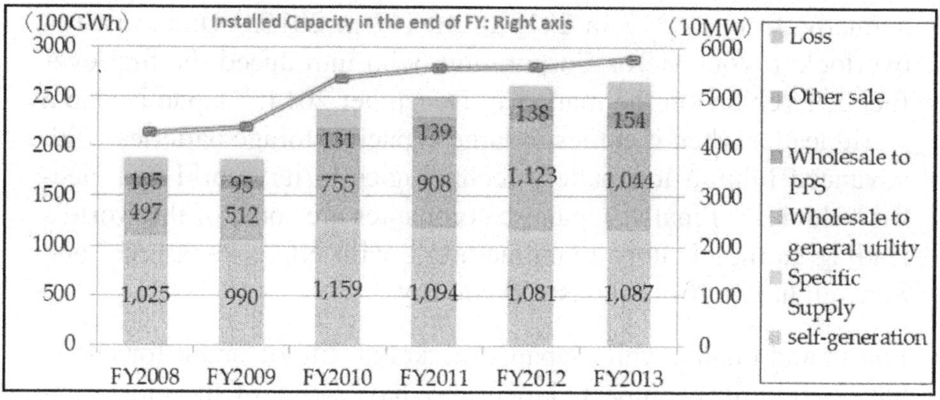

Figure 3. Generated Electricity and Installed Capacity of Self-Generation in Japan.[14]

Second, Japan still has not introduced very much RES at all. If large-scale hydro-electric generators are excepted, RES account for only 1.7% of Japan's electricity generated,[15] well behind most other developed countries. Two of the largest contributing factors are the fact that renewables received a very low policy priority and the fact that it is very difficult for them to connect to the grid under the vertically integrated system. However, Japan possesses even greater resource potential than Germany,[16] one of the leading

14. Total installed capacity of Japan's 10 general electric companies is 209.32GW. Agency for Natural Resources and Energy, "Electricity Survey Statistics," March 31, 2014.

15. Renewable energy electricity generation was 16.94TWh (Ibid).

16. According to a 2011 report of a government cost verification committee, under certain developmental conditions, Japan's renewable energies were estimat-

countries in renewables in the world. Japan also has many leading manufacturers of solar panels, wind turbines, and geothermal generators. Accordingly, a large introduction of renewables would have a significant impact in terms of Japan's industrial policy, as well.

Third, although fuel cells and large capacity storage batteries are still in their initial stages globally, Japan is positioned at the forefront of the field. Japan's sales of household-use fuel cells have expanded annually, from 24,517 in 2012 to 40,447 in 2015.[17] One must not overlook Toyota Motor Corporation, who introduced the first ever fuel cell vehicle on the market in December 2014.[18] Japan is also a world leader when it comes to large capacity storage batteries, with advanced lithium ion battery technologies in its hybrid and electric vehicles.[19] Finally, Japanese companies are some of the world's leading storage battery manufacturers, with NEC, Panasonic, and Sony all hailing from the island country.

Fourth and finally, while Japan's market is still too small for DR to have a significant impact, Japan may have some of the most conscientious consumers in the world. When supply-demand was tight

---

ed as follows: Solar 91GW, on-shore wind 150GW, geothermal 4.3GW. National Policy Unit, Energy and Environmental Council, "Cost Verification Committee Report," presented at the Energy and Environment Conference, December 19, 2011, Japanese version available from *www.cas.go.jp/jp/seisaku/npu/policy09/pdf/20111221/hokoku.pdf.*

17. Third, although fuel cells and large capacity storage batteries are still in their initial stages globally, Japan is positioned at the forefront of the field. Japan's sales of household-use fuel cells have expanded annually, from 24,517 in 2012 to 40,447 in 2015.

18. Toyota, "Toyota Opens the Door and Invites the Industry to the Hydrogen Future," *Toyota-USA Newsroom*, January 5, 2015, available from *corporatenews.pressroom.toyota.com/releases/toyota+fuel+cell+patents+ces+2015.htm.*

19. According to the Next Generation Vehicle Promotion Center website, vehicle retention numbers in 2012 were: PHV 17,281; EV 38,707; and hybrid 2,852,105. See "Number of Electric Vehicles Owned (Point Estimate)," Next Generation Vehicle Promotion Center, available from *www.cev-pc.or.jp/english/archive/white-papers/owned.html.*

following Fukushima, DR was thought to be an important solution, and DR aggregators such as NTT Facilities and Eneres worked together with power companies to roll out a program. Foreign companies such as America's EnerNOC and France's Energy Pool have also entered Japan's market. Unfortunately, under the current vertically integrated system, market mechanisms don't work and there are still no real-time balancing or capacity markets; electricity system reform is needed before these things can happen.

Energy Policy after Fukushima and Future of Distributed Energy

As we have seen above, Japan has plenty of potential for developing the distributed electricity system. However, DES has received very little policy backing to date and consequently has remained ineffective, and Japan's centralized electricity system of the past remains firmly sustained. The fate of DES in Japan therefore hangs on an array of policies. As the title of this paper implies, I will now examine whether or not the electricity system reform after Fukushima and other energy policies are truly aiming for DES.

First of all, Japan's policy position on DES remains unclear. In April 2014, the Abe administration published the first Basic Energy Plan since Fukushima, but it was very broad and difficult to understand. RES are described as "an important domestic low-carbon source of energy,"[20] but at the same time the document says of the future of the feed-in tariff program begun in 2012 that it would be "generally reviewed" from the viewpoint of "minimizing the burden on citizens."[21] Based on this plan, the tariff to purchase electricity produced from RES was continuously reduced, and the Feed-in Law was revised in 2016 so that auction to large-scale solar photovoltaic (PV) would be introduced in 2017.

---

20. In the original proposal in December 2013, renewable energy was not described by the adjective "important" and was not highly evaluated. The adjective was added later, following a party conference.

21. Government of Japan, *Basic Energy Plan*, April, 2014.

Second, the Basic Energy Plan still assigns a key role to centralized power sources. Nuclear power plants, which had all been shut down by September 2013, are described as "important base load power sources." Because it is difficult for nuclear power plants to maintain their businesses due to stricter safety standards put in place after the Fukushima nuclear disaster, the government is considering new ways of supporting them, such as limiting liability in the case of a severe accident, covering the costs of decommissioning reactors, and the implementation of a Contracts for Difference program as in Great Britain. Coal is also given a place as "an important fuel for base load power sources." After Fukushima, the environmental assessment standards for coal-fired thermal power plants were relaxed, and 5GW worth of new facilities is planned for the early 2020s.

As it happens, the Basic Energy Plan has a section titled "Expanding the Distributed Energy System." However, no clear definition is given, and judging from the context, it seems to mean "distributed energies such as renewables and cogeneration" will be managed "in communities of a certain size" "using IT and storage battery technologies."[22] Essentially, this means that the centralized electricity system will be preserved and a distributed system will be used in a supplementary role in certain areas, thereby absorbing the variability of RES.

Third and finally, although DES requires electricity reform, the detailed reform plan has many compromises as stated above, and it will not be implemented quickly. For example, not only was the grid unbundling watered down to a very weak structural measure, i.e., legal unbundling, but it is set to be implemented no later than 2020. Competition policy in power generation is also not getting the reforms it needs. Power generation facilities will not be bought back as they were in the United States, nor will dominant generation companies be forced to sell their wares on the wholesale market. Furthermore, it is general practice to establish an independent regulatory body when liberalizing the electricity market, but the gov-

22. Ibid.

ernment seems set to establish a minimally independent body, the so-called article-8 council, within the Ministry of Economy, Trade, and Industry. While the government touts electricity system reform as one of the pillars of its growth strategy, it seems to be taking pity on the financially struggling power companies and moving forward very slowly and with many compromises.

Outlook on Distributed Energy Sources in East Asia

Industrial Infrastructure for Distributed Energy Sources in Non-Western Countries

In Western developed countries, there has been a trend from centralized energy to distributed energy, from centralized electricity systems to distributed ones. Conversely, rapidly growing developing countries still rely heavily on coal-fired thermal and nuclear power generation. Many of these countries have not liberalized their electricity markets and still have state-run power companies with vertical integration. Without a shift in their energy policies and electricity systems, it is doubtful that DES will spread.

Another possibility for accelerating DES is the precipitous drop in costs brought about by a disruptive technological innovation. For example, in the 1990s developing countries did not have many landline telephones, but the IT revolution led to the sudden prevalence of cellular telephones. The more complete a country's centralized electricity system is, the more difficult it is to realize structural reform; by this measure, it is entirely possible for the so-called "leap frogging phenomenon" to occur in the electricity field in developing countries.

Japan's neighboring countries also have no small amount of potential when it comes to DES. China has the most installed capacity for wind power in the world (Figure 4), and in 2013 set a record for the most solar panel capacity installed in a single year (11.3GW) and jumped to number two in the world in terms of cumulative capacity (18.3GW).[23] In terms of industrial infrastructure, China has world-leading solar panel manufacturers, and 60% of all photovoltaic (PV) cells made in 2012 globally were made in China.[24] South Korea is lacking in renewable energy resources, but its storage battery industry rivals even Japan's.

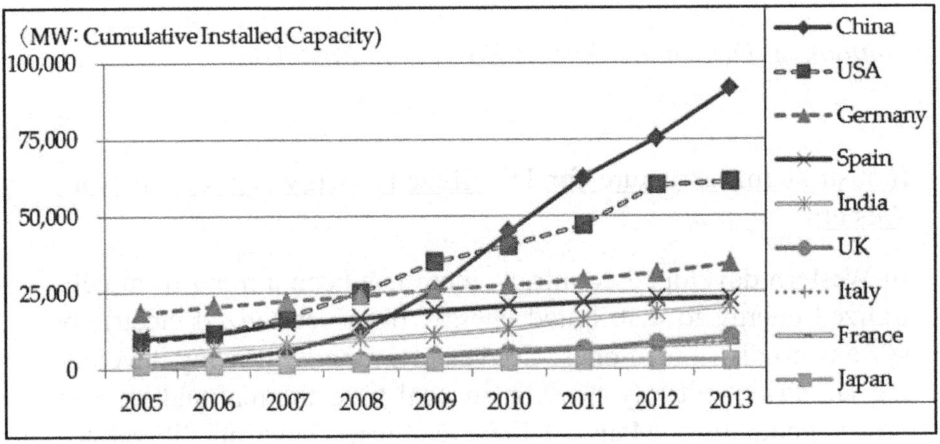

Figure 4. Installed Capacity of Wind Power Generation in Leading Countries.[25]

The industrial infrastructure of related products in East Asia therefore has plenty of potential, but those countries are lagging behind in terms of policy. First of all, electricity liberalization has made little progress, and Japan and South Korea have had only limited success in introducing renewable energies due in part to issues with opening

23. #1 was Germany with 35.5GW. Japan was #4 with 13.6GW. See, Gaëtan Masson, *PVPS Report: Snapshot of Global PV 1992-2013*, Report IEA-PVPS T1-24:2014, Paris, International Energy Agency Photovoltaics Power System Program, 2014.

24. Ibid.

25. The chart shows the Top 8 countries in 2013, and Japan, which was #16. Global Wind Energy Council, *Global Wind Report: Annual Market Update 2013*, available from *www.gwec.net/wp-content/uploads/2014/04/GWEC-Global-Wind-Report_9-April-2014.pdf*.

up the transmission grid. An inability to use market mechanisms means that DR cannot be used effectively either. On the other hand, because liberalization is at a standstill, nuclear power is in a very advantageous position. In fact, China and South Korea have stuck to their plans to build many new nuclear power plants even after Fukushima. The fact that China still relies on coal-fired thermal generation for 70% of its power generation tells us that DES still has a long way to go in these countries.

Will East Asia Forge International Grid Linkages?

When it comes to electricity systems, there are also significant obstacles in terms of grid operation. Both Japan and South Korea have independent grids with no international linkages. This means that there is a limit to the range of their grid operations, which presents a problem when it comes to counteracting the variability of electricity provided by RES. These countries contrast starkly with European nations, which have deeply integrated their transmission grids and electricity markets and are standardizing their energy policies, including those regarding renewables.

Is there no hope, then, for international linkages between grids in East Asia? The author believes there to be ample potential, both technologically and economically. For example, Fukuoka city in Japan and Busan city in South Korea are separated by approximately 220km of sea. The longest undersea transmission line in the world, NorNed, which connects Norway to the Netherlands, is 580km long. Hokkaido and Sakhalin, Russia are only 40km apart. Japan could most certainly form international grid linkages with other East Asian countries.

If grids were internationally linked, differences in electricity prices would likely result in brokerages engaging in much export and import of electricity. Japan's price for electricity is more than twice South Korea's; with such a large disparity, brokerages would thrive. Furthermore, if both countries were to introduce large amounts of renewable energy, electricity import and export would play a large role in ensuring the stability of both grids. If South Korea's grid were further linked to North Korea, China, and even Mongolia, the stabilizing effect would be even greater.

Softbank, one of Japan's leading telecommunications companies, proposed the idea to connect all of Asia with an ultra-high-voltage transmission grid, build enormous wind and solar farms in Mongolia, and thereby supply electricity to Japan, designating it the "Asia Super Grid Initiative." This is not a mere daydream, however, as Softbank is working together with Mitsui & Co., a large trading company, and Inter RAO, a large Russian power company, to investigate the business feasibility of importing electricity from Sakhalin to Japan.[26]

While the private sector has begun moving to create international grid linkages, the Japanese government's reaction has been slow due to the souring diplomatic relations with neighboring countries. Relying on each other for energy requires a certain level of trust between two countries, and unfortunately, such trust is not being developed in East Asia, particularly between Japan and other leading countries such as China and South Korea. In order to build an international transmission grid and make proper use of it, market integration policies must also be in alignment, but the current state of things in East Asia does not allow very much room at all for cooperation on policy or anything else.

26. "Power imported from Russia concept Softbank and Mitsui & Co. Ltd." Nihon Keizai Shimbun, February 24, 2013, Japanese version available from *www.nikkei.com/article/DGXNASDD2302L_T20C13A2MM8000/*.

Conclusion

In conclusion, the outlook for DES in Japan is not very bright. The government asserts that it will decisively undertake electricity system reform, but its progress is incremental, the order of priority among renewables, cogeneration, nuclear, and coal as power sources is unclear, and the energy policy seems directionless. To give things a more favorable interpretation, one might say that the government is trying to give itself policy flexibility to respond to the present unclear energy environment, but their hemming and hawing is doing nothing more than to put off the difficult decisions, and there is a distinct possibility that Japan will be stuck on the fence for good. The uncertain policy environment may cost the energy industry a lot of yen in private sector investment.

Even within the larger framework of East Asia, DES's lot does not change all that much. Japan, China, and South Korea all have the physical and industrial potential to implement DES, but they all assign it a very low policy priority. Cross-border competition, and sometimes cooperation, helped spur the development of a distributed electricity system in Europe, but East Asia is far from being able to follow that example.

If any momentum towards DES is to be had, it will be from the direction of market movements and innovative developments. Market trends are much better able to cross international borders than policies. Additionally, East Asian countries are susceptible to the policies and market trends of Western countries. If Western energy companies enter East Asian markets, they may shift momentum towards DES. The future will reveal whether the potential for DES in East Asia will be realized or not.

CHAPTER 4

Natural Gas Price in Asia: What to Expect[1]

Kenneth B. Medlock III

Introduction

During the past decade, innovative new techniques involving the use of horizontal drilling with hydraulic fracturing have resulted in the rapid growth in production of natural gas from shale in the United States. This has transformed the North American gas market and caused ripple effects around the entire world. To begin, the successes in the United States in shale gas and tight oil production have triggered tremendous interest in unlocking similar potentials in other identified shale resources around the world. While an array of above-ground factors will limit the pace of development in many places relative to what has been witnessed in the United States, the interest is strong, and investment capital is seeking to make the opportunity real.

The transformative impacts of shale development extend beyond the interest in developing such resources outside the United States. A little over a decade ago, steady production declines were expected in the North American market. This, in turn, signaled an increasing reliance on imported supplies of liquefied natural gas (LNG). In anticipation of rising demand for LNG from the United States, developers around the world began investing in expanding LNG export capability, concomitant with investments in regasification being

1. This paper is also available from *bakerinstitute.org/media/files/Research/ac817540/CES-pub-NaturalGasPriceAsia-021814.pdf*.

made in the United States. But rapid growth in shale gas production in the United States has rendered many of these investments obsolete. More importantly, as discussed in Medlock, Jaffe, and Hartley (2011),[2] it forced those LNG supplies that had been developed with the United States as a target market to seek new ports of destination. For a while, this displacement effect put substantial downward pressure on spot prices in both Europe and Asia. This was exacerbated by the economic malaise that afflicted global markets in the latter part of last decade. However, prices eventually rose as the disaster at Fukushima turned the global gas market around by generating a wholly unexpected demand shock. This, in turn, allowed those displaced LNG supplies to find a new home. Arguably, this effective "capacity release" from the United States has helped keep prices in Asia from rising even more than they have.

Another impact of rising shale gas production in the United States has been the growing interest in exporting LNG. Certainly, the rapid production growth has contributed to lower domestic natural gas prices, which dipped below $2 per thousand cubic feet (mcf) in April 2012 and, prior to the demand pull created by the cold winters that currently grip the United States, hovered in the $3 per mcf range for an extended period of time. Low prices led to greater use of natural gas in power generation through substitution opportunities with coal, a revival of industrial and petrochemical demands, and growing interest in expanding natural gas use in transportation. Low U.S. prices relative to the prices in Europe and Asia have also triggered interest in developing LNG export capability to capture the profitable arbitrage opportunity that currently exists.

As discussed in a paper from 2012, when considering international natural gas trade, it is important to recognize that the issue is indeed *international*.[3] Thus, we must not only consider what is happen-

2. Kenneth B. Medlock III, Amy Myers Jaffe, and Peter Hartley, "Shale Gas and U.S. National Security," James A. Baker III Institute for Public Policy, Rice University, July 2011, available from *bakerinstitute.org/files/496/*.

3. Kenneth B. Medlock III, "U.S. LNG Exports: Truth and Consequence,"

ing in North America; we must also consider what is happening abroad. Only then can we begin to analyze what the future may hold for natural gas pricing and trade around the world. In fact, as natural gas becomes an increasingly fungible commodity, which would be the case as the volume of global natural gas trade increases, the pricing paradigm of oil-indexation will come under increasing pressure.

There are several key factors that determine the impact of LNG trade on prices in all markets, including but not limited to (1) the relative long-run elasticity of domestic and foreign supply; (2) the relative long-run elasticity of domestic and foreign demand; (3) the role of short-term capacity constraints as they will be impacted by the introduction of trade; and (4) the cost of developing and utilizing export capacity.

Identifying unexpected, transitory events is crucial to characterizing the current natural gas market. In general, unexpected changes in demand can create transitory price movements, particularly when supply cannot react quickly. This is seen even in the continental North American market when extreme cold grips certain regions and drives up local demand in excess of pipeline delivery capability. But, fortunately for consumers in the United States, these regional price shocks do not last long, since weather-related demand shocks are short-lived, and the depth of the U.S. market provides substantial liquidity through which price differences are quickly arbitraged. As argued below, the strength in Asian LNG price coincides with the unexpected increase in demand that occurred in the wake of the disaster at Fukushima on March 11, 2011, and the subsequent shutdown of the entire Japanese nuclear power generation fleet. This demand shock created tightness in the LNG market that dramatically influenced the spot price of LNG in Asia. However, as new capacity emerges to deliver supplies into the Pacific Basin market and/or nuclear capacity in Japan is re-started, the short-term tightness will be alleviated, which should have a countervailing effect on price.

James A. Baker III Institute for Public Policy, Rice University, August 10, 2012, available from *bakerinstitute.org/files/842/*.

This chapter begins with a discussion of U.S. LNG exports as a classic problem in international trade, which makes it possible to assess the likely future of price in the Asian LNG market couched in the context of *international* trade. Then, results from the Rice World Gas Trade Model (RWGTM)[4] are presented in order to highlight the effect of a deepening global gas market on the price of natural gas in Asia. This, of course, has implications for the attractiveness of other energy options, such as expanded use of coal, oil, and even nuclear power in the electricity generation mix in the long-term.

Asian LNG Price in a Basic Trade Paradigm

Begin by considering the issue of the future of Asian gas pricing in the context of an international trade model. Figure 1 shows a domestic and a foreign market in an autarkic equilibrium, that is, one in which there is no trade between the two markets. The supply-demand equilibrium in the domestic market yields a price below that of the supply-demand equilibrium in the foreign market. If the spread between the two prices, which would be denoted as $P^{*f} - P^{*d}$ in Figure 1, exceeds the cost of liquefaction, shipping, and regasification, it leads to an "arbitrage opportunity" that can be exploited through trade.

If the price spread between the foreign and domestic markets is large enough to cover the costs associated with trade between the two markets, then there is sufficient economic incentive for trade. Figure 2 displays the impact on price in both markets if trade occurs such that the domestic market sees exports of χ and the foreign market sees imports of m, where $\chi = m$. Price rises in the domestic market and falls in the foreign market. If not constrained by policy,

4. The Rice World Gas Trade Model is developed on the MarketBuilder software platform from Deloitte MarketPoint. It is available to the Baker Institute through an academic license agreement. All model characterizations of infrastructure as well as all data inputs for demand, supply, and infrastructure and development costs have been independently developed by Kenneth B. Medlock III and Peter Hartley at the Center for Energy Studies of Rice University's Baker Institute.

such as quotas or tariffs, an equilibrium will be reached in which price in the two markets differs only by the cost associated with trade, indicated as τ in Figure 2.[5]

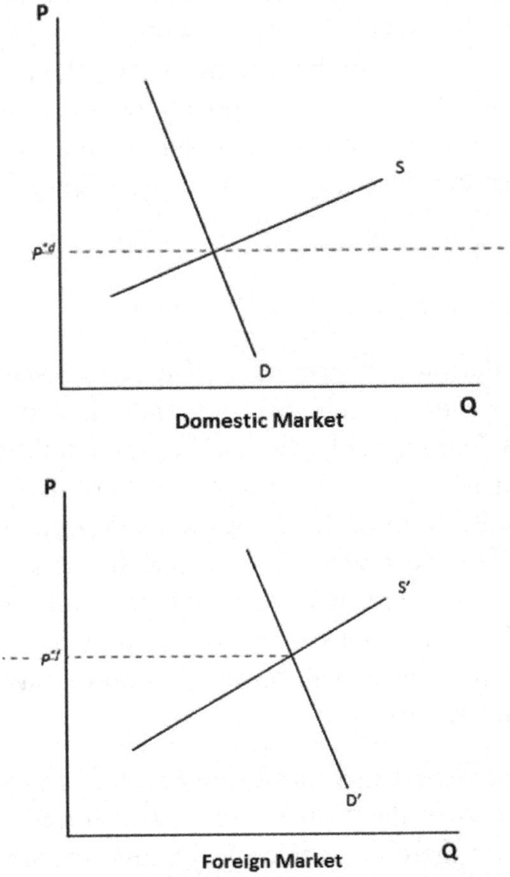

Figure 1. Domestic and Foreign Market with No Trade.

5. Henceforth, the sum of liquefaction and shipping is referred to simply as "transport costs" or the cost of trade.

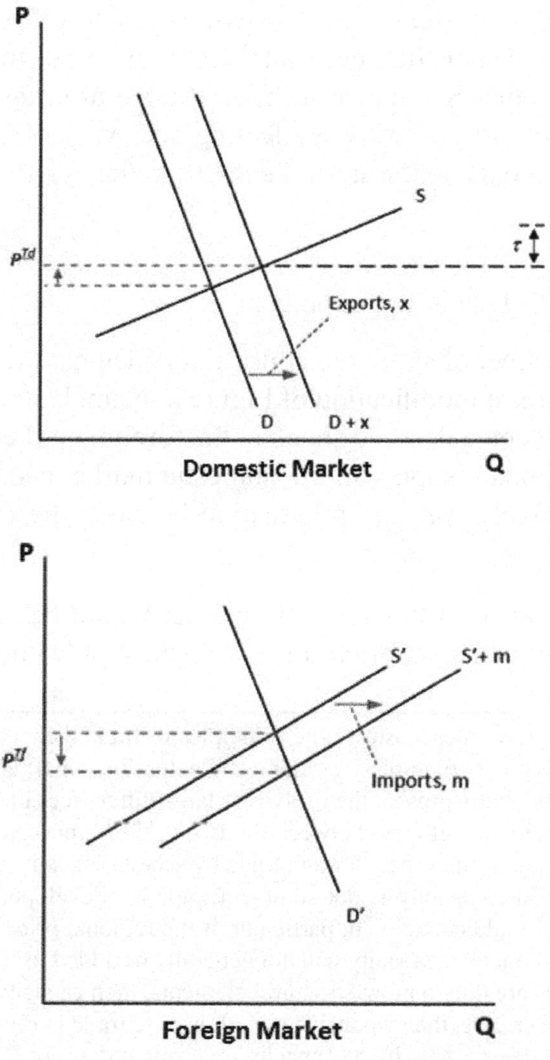

Figure 2. Domestic and Foreign Market with Trade.

In general, theory tells us that in the long run, as market participants seize arbitrage opportunities, prices will adjust to reflect the cost of arbitrage. In the short run, demand shocks and other transitory factors may present what appear to be profitable arbitrage opportunities, but these will generally be fleeting and will not support large capital investments necessary to facilitate trade.[6]

From the Short Run to the Long Run

Framing the effect of short-run constraints on apparent trade opportunities requires a modification of Figure 2. Namely, the impact of a sudden, unexpected demand shock in the foreign market alongside a relatively abundant supply in the domestic market can be captured. This is effectively the current situation between the United States and Asia.

Supply abundance in the domestic market would be captured by a highly price responsive, or elastic, domestic supply curve.[7] Accord-

---

6. As an illustrative microcosm of the principles of trade theory in practice, we can consider what occurs in the U.S. domestic natural gas market. Arbitrage opportunities occasionally present themselves as large differences in regional prices. If pipeline capacity is sufficient between the two regions, marketers will quickly eliminate the pricing difference through trade by scheduling shipments across the pipeline. If pipeline capacity is not sufficient, pipeline developers will evaluate the opportunity to add capacity. In particular, if the regional price differences are due to short-term factors, capacity will not generally be added. But, if the regional price differences are due to more structural elements, then capacity will generally be added. In either case, the responsiveness of price to trade in both regional markets is a critical determinant to the capacity investment decision.

7. Note that we do not focus on the elasticities of demand because in both the domestic and foreign markets, demand is being driven by growth in power generation requirements. Given the lack of technological differences, the availability of competing fuels and the fact that demand for natural gas is relatively own-price inelastic in all major end-use markets—generally varying between 0.15 and 0.3 according to Baker Institute analysis—we focus instead here on the relative elasticities of supply. In fact, allowing for variable elasticities of demand will tend to reinforce the results herein.

ing to the Baker Institute study, "Shale Gas and U.S. National Security," the elasticity of supply in the United States is fairly high for prices between $4 and $6 per mcf.[8] Importantly, this elasticity of supply so characterized for the United States is a *long-run* elasticity. Long-run elasticity is used when considering the price impact of *expected* events, such as the opening of an LNG export terminal, because it is a more appropriate representation of supply responsiveness. Producers know the additional market "demand" in the form of exports is coming because the opening of the terminal is public knowledge. Thus, the additional demand for U.S.-produced natural gas in the form of exports abroad should not be treated as a surprise, which is what using a short-run elasticity in this instance effectively does.

In the foreign market, one must first recognize the impact of an unexpected demand shock, such as in the wake of Fukushima. Following the tsunami, the resulting nuclear accident sparked concerns that ultimately led to the closure of all of Japan's nuclear power generating capacity and resulted in a dramatic increase in Japanese demand for LNG. This increase in demand is captured in Figure 3 with a move from *D* to *D'*. It must also be recognized in the short term that there is a constraint on the ability to deliver supplies (where the constraint is represented by the vertical portion of the supply curve, *S*). It takes time to develop new supply capacity and a *sudden*, unexpected increase in demand can result in binding short-term capacity constraints. The situation can be especially pronounced when storage capacity is lacking and/or there is an inability to physically hedge against unexpected events.

The impact of the sudden demand increase in Japan led to a large increase in the spot price of LNG in Asia. In fact, the Platts Japan/Korea Marker price, which is the benchmark daily assessment of the spot price for cargoes of LNG delivered ex ship to Japan or

8. Indeed, that study indicates the price elasticity of supply, post-shale, has risen over five-fold, from 0.29 to 1.52. "Shale Gas and U.S. National Security" was sponsored by the Office of International Policy and Affairs of the U.S. Department of Energy.

Korea, increased by over $2 per mcf in the week following the incident at Fukushima and continued to climb in the following months, concurrently with the shutdown of all of Japan's nuclear power generation plants.

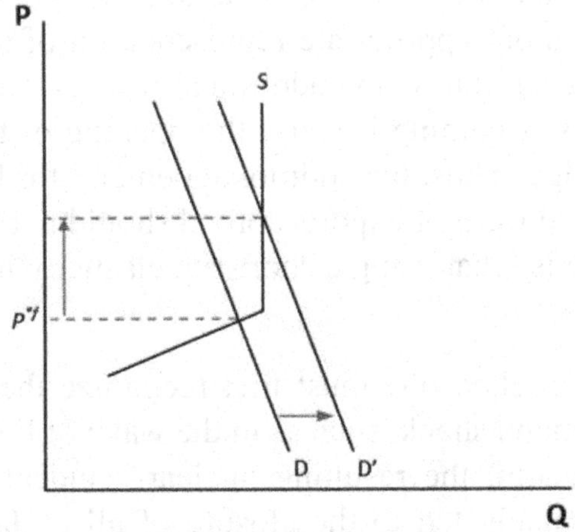

Figure 3. The "Foreign" Market for Natural Gas.

Figure 4 indicates the monthly average prices of natural gas at the U.S. Henry Hub (HH), the UK National Balancing Point (NBP), the Platts Japan/Korea Marker (JKM), and a representative crude-indexed LNG price from 2009 to 2014. Following the nuclear incident at Fukushima and the subsequent shutdown of all of Japan's nuclear reactors, the JKM price ventured into unprecedented territory. In fact, as can be seen in Figure 4, JKM climbed markedly relative to both NBP and HH after Fukushima, with the spot price approaching oil-indexed parity. It is also worth noting that the standard deviation of the spread of daily prices between NBP and JKM is over two times higher post-Fukushima. Higher price and higher price volatility are both indicators of the realization of a constraint on the ability to deliver supply to the Asian market.

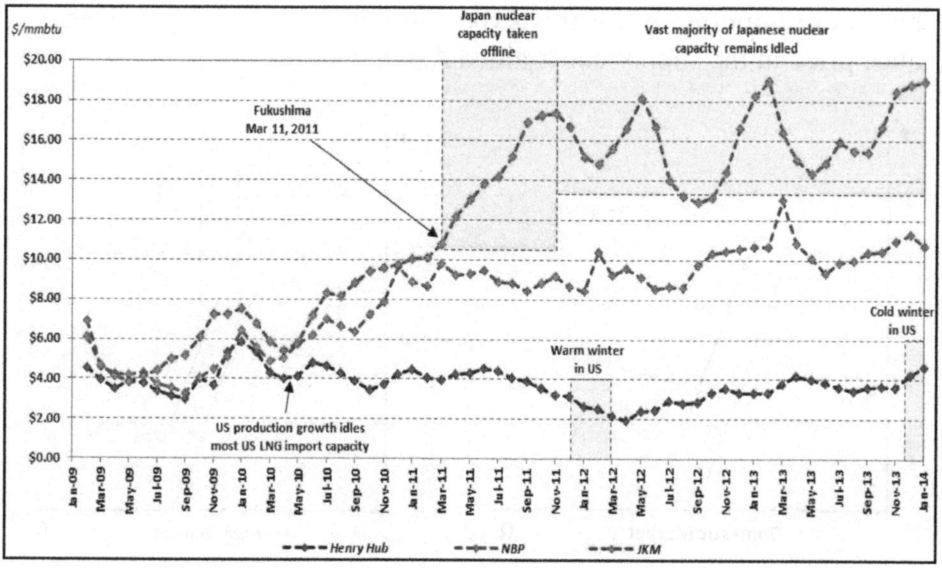

Figure 4. Global Marker Prices Annotated (Monthly, Feb 2009 - Jan 2014).[9]

In general, if we add supply to a supply-constrained market, the price in that market will fall precipitously, all else equal. In the case of the Asian natural gas market, supply will be added—whether in the form of LNG exports from the United States or from other sources of supply via pipeline or LNG—because the current price spread is more than adequate to encourage such a response. Figure 5 illustrates that the price paid by LNG importers in Asia will fall substantially when the short-term constraint on the ability to deliver supply is abated. If the "foreign" market in Figure 5 is taken to represent the Asian gas market, the addition of LNG, or any other supplies for that matter, will have a large effect on Asian LNG price, even if the demand shock, Δ, is not reversed.

If Japan's nuclear power generation fleet is brought back into commission, the downward movement in price would only be exacerbated. Of course, the resultant downward pressure on price in Asia will ultimately determine the amount of trade that actually occurs,

9. Platts, U.S. Energy Information Administration.

but, more importantly for this exposition, any new supply will reduce price in the supply constrained Asian market.

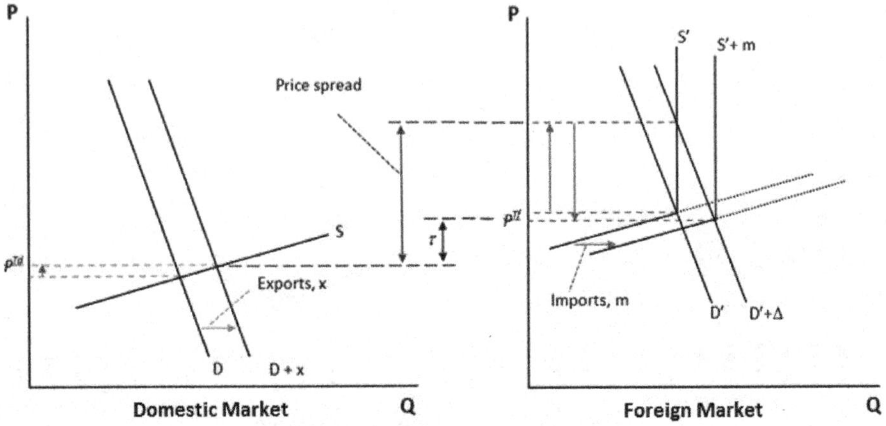

Figure 5. The Impact of Trade on the Asian Price.

Importantly, the *relative* elasticities of supply and demand will determine the extent to which prices change in each of the traded markets. In particular, if the market abroad is short-term supply-constrained (meaning supply is inelastic) and the domestic market is supply-elastic, then the majority of the price movement that will occur when trade is introduced will be abroad. To the extent that the current price difference between Asia and the rest of the world is due to transitory factors that result in the realization of supply constraints, or short-term inelasticity, the current price differential will only persist until there is some offsetting influence. Thus, the pre-Fukushima pricing relationship between JKM and NBP can be expected to re-emerge as (1) new LNG supply capability is developed; (2) new sources of pipeline supply to Asia are developed; and (3) Japan's nuclear reactors are restarted.

One final but important point regarding the effects of increased trade pertains to market liquidity. As the Asian gas market deepens with new sources' supply being introduced, an ability to arbitrage regional price differentials will push prices into relative ranges defined by transportation costs. It is the lack of capability to arbitrage current regional price differences, or lack of physical liquidity, that allows regional prices to drift apart dramatically. If the United States develops export capability, an additional arbitrage mechanism will be introduced, and, all else equal, this will force a shift in the relative prices of gas in markets around the world to long-run equilibrium differentials that are defined by transportation costs and currency values.[10] This will impact more than just the trade at the margin or spot price. Greater liquidity will also reduce the willingness of consumers to pay prices above marginal cost for supplies.[11] This will, in turn, have direct implications for oil indexation as a pricing paradigm in the Asian market.

Implications for Market Liquidity and the Pricing Paradigm

When a profitable arbitrage opportunity exists, excessive rents can only persist in the long run if suppliers—regardless of location—are willing to forego the fixed costs associated with expansion to capitalize on the opportunity. With regard to the Asian LNG price, this means the current differentials can only persist if U.S. producers are willing to forego future LNG export capacity expansion in the face of large rents, or if they are prevented through policy action from doing so. It would also be necessary that all other supplies—East African LNG, Russian pipeline gas, etc.—either be prevented from delivering supply to Asia or be more costly to deliver to Asian consumers than U.S. supply. In the latter case, the current price in Asia could be viewed as accurately reflecting the marginal cost of non-U.S. resources delivered to Asia. Moreover, in this case, only

10. Although not explored in this exposition, currency values are important to determining price differentials because natural gas is not delivered to end-users in a common currency denomination. See Kenneth B. Medlock III, "U.S. LNG Exports: Truth and Consequence" for more on this issue.

11. Peter Hartley, "The Geopolitics of Natural Gas: The Future of Long-Term LNG Contracts," James A. Baker III Institute for Public Policy, October 2013, available from *bakerinstitute.org/research/future-long-term-lng-contracts/*.

U.S. suppliers could realize rents by serving the Asian market.

It is highly unlikely that such a specific set of restrictive conditions exists. For one, given the current push to develop LNG export terminals in the United States, it is evident that developers will seek to capture profits associated with LNG exports to Asia in spite of the high upfront costs. Moreover, the quantity and cost of supply around the world indicate that there is a large amount of natural gas that could be delivered to Asia at well below the current price. This is particularly true in regions with high quantities of associated liquids—such as in currently producing areas in the Middle East and North and West Africa—and is likely true in regions with newly identified resources that are not currently exporting LNG or pipeline supplies to Asia—such as East Africa, South America, and Russia. Already, developers in those regions are directing development activities toward serving Asian demand. To varying extents, there are policies and institutional factors that may inhibit development in some regions, but these are not universal; thus, a supply response to the current high price environment can be reasonably expected.

Given the nature of the supply constraint in Asia, one should expect the emergence of multiple competing opportunities to provide natural gas supplies to Asia. Examples could include development of shale resources in China; pipeline options from Russia, Central Asia, and South Asia; and LNG supplies from Australia, East Africa, the Middle East, and North America. As detailed above, the current arbitrage opportunity is being aided by short-run *inelasticity* of supply to Asia, but this cannot be expected to persist in the long run because of the number of different potential supply options.

Analysis at the Baker Institute indicates that Asian natural gas prices will likely soften relative to their current levels. This reflects:

- Long-term shale developments in places such as China and Australia will become commercially attractive in price environments in excess of $10 per mcf. Indeed, wide-scale success in shale development in China could be as transformative to the global gas market as the successes in North America have been over the past decade.

- The development of pipeline supplies from Russia, Central Asia, and South Asia to China will displace the need for LNG into China, all else equal. This frees up those LNG supplies for consumers in Korea and Japan. Thus, pipeline supplies serve as another point of competition for LNG long-term, particularly in developing continental markets.
- Furthermore, if the Japanese nuclear situation reverses and we see the reactivation of the nuclear power generation fleet across the country, the currently constrained market equilibrium will be alleviated just as if new supplies were brought to bear. Arguably, the current price of natural gas delivered to Japan is a catalyst in this direction. If the nuclear generation fleet is reactivated, it will serve to accelerate the pace at which prices adjust downward, particularly if the demand for LNG falls just as new supplies are made available. In essence, the market would be forced into an excess capacity situation and price would be driven down as competition for customers would be intensified.

All of these factors support the thesis that growth in trade in the Asian market will foster market deepening and increased liquidity. As demonstrated in Hartley (2013), this will, in turn, support a movement away from the traditional pricing paradigm of long-term, oil-linked contracts. This follows because growth in market liquidity reduces the risk of being able to secure supplies. As long as oil price remains above the price of various new natural gas supply options, the willingness to pay above marginal cost due to security-of-supply concerns is diminished (see Figure 6). In fact, Brito and Hartley (2007)[12] show that as physical liquidity increas-

12. Peter Hartley with Dagobert Brito, "Expectations and the Evolving World Gas Market," *Energy Journal* 28, no. 1, 2007, pp. 1-24.

es, the ability of any single supplier to set a price above marginal cost is diminished. This, all else equal, will mean oil-indexation is likely to lose its prominence. Importantly, there is no guarantee that movement away from oil-indexation will result in natural gas prices falling long-term relative to crude oil; rather, a lack of oil-indexation should only mean that gas will be priced according to its marginal cost. It should be noted that *contract* prices can be substantially different from *spot* prices. As indicated in Figure 6, in this case, one can think of contracted supplies as being infra-marginal and the result of price discrimination.

In general, for a firm to be able to price discriminate, (1) it must be able to distinguish consumers and prevent resale, and (2) its consumers must have identifiable and different elasticities of demand. An increased ability to trade (i.e., increased physical liquidity) violates condition (1). This is more likely to happen as the supply curve in Figure 6 becomes more elastic (flatter). It is also prone to happen in a liberalized market where trading is unfettered because price signals communicate arbitrage opportunities that can be captured. This promotes entry and, to the extent that hubs develop, financial liquidity. Once that occurs, the means to use capital markets to underwrite physical transactions increases and liquidity grows, thus making it difficult to price discriminate. As liquidity increases along with the long-term elasticity of supply, the rents associated with infra-marginal contracts also increase, which, in turn, triggers calls for renegotiations of terms between producers and consumers.

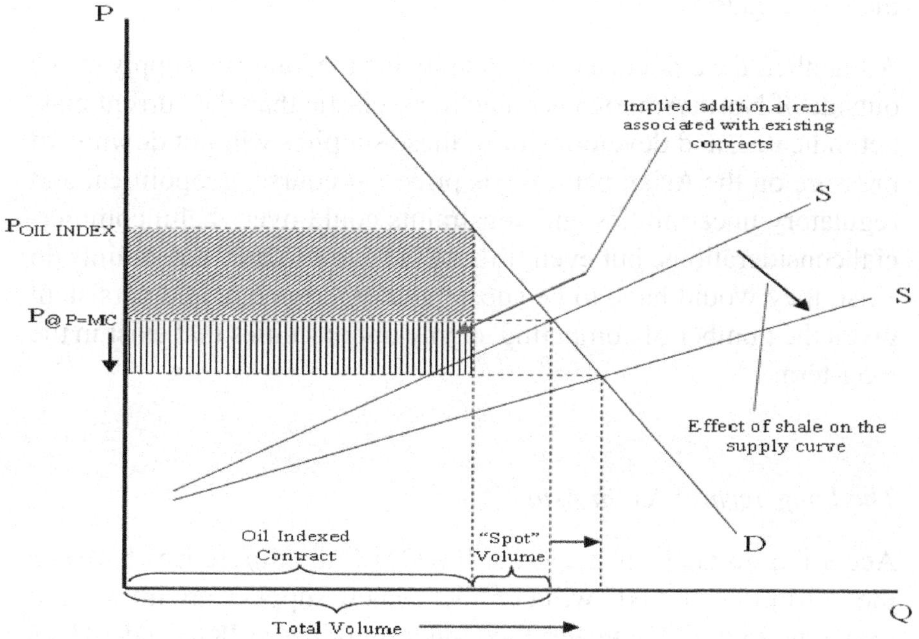

Figure 6. The Supply Curve Effect of More Elastic Supply and Implications for Price.

Relatively high prices in Asia have already encouraged producers to begin investigating the opportunity for profitable shale development in those markets. The initial forays into shale in Asia and other regions have proven to be more costly than what has been experienced in the United States. Much of this is due to lack of a well-developed upstream service sector, which means there is limited availability of equipment and personnel. However, as those sectors develop, encouraged by sustained demand pull from increasing upstream activity, the current high costs should prove transitory. Accordingly, the prospects for long-term shale development in places such as China, Australia, and Argentina—the latter two of which could serve the Pacific Basin via LNG—look promising. In fact, a Baker Institute analysis revealed that shale gas developments in China could be equally as game-changing over the next couple of decades as shale gas developments in North America have been in the last decade.[13]

Altogether, the evidence is substantial that the long-run supply curve outside of North America is much more elastic than the current market indicates, and development of these supplies will put downward pressure on the Asian natural gas price. Of course, geopolitical and regulatory uncertainties and constraints could overwhelm commercial considerations, but even if these "above-ground" constraints do exist, they would have to be substantial, widespread, and persistent given the number of competing supply opportunities that exist in the long-term.

The Long Term Price in Asia

According to analysis using the RWGTM, the differences between the JKM price and HH will contract as new supplies are introduced into Asia. Figure 7 indicates the long term prices at JKM and HH, as well as the price spread juxtaposed against the cost of trade between the two markets.

---

13. Kenneth B. Medlock III and Peter R. Hartley, "The Rise of China and Its Energy Implications: Quantitative Analysis of Scenarios for Chinese Domestic Unconventional Natural Gas Resources and Their Role in Global LNG Markets," James A. Baker III Institute for Public Policy, December 2, 2011, available from *bakerinstitute.org/files/529/*. This study revealed that shale gas development in China could be as game-changing in the coming years as shale gas developments in North America have been in the last decade.

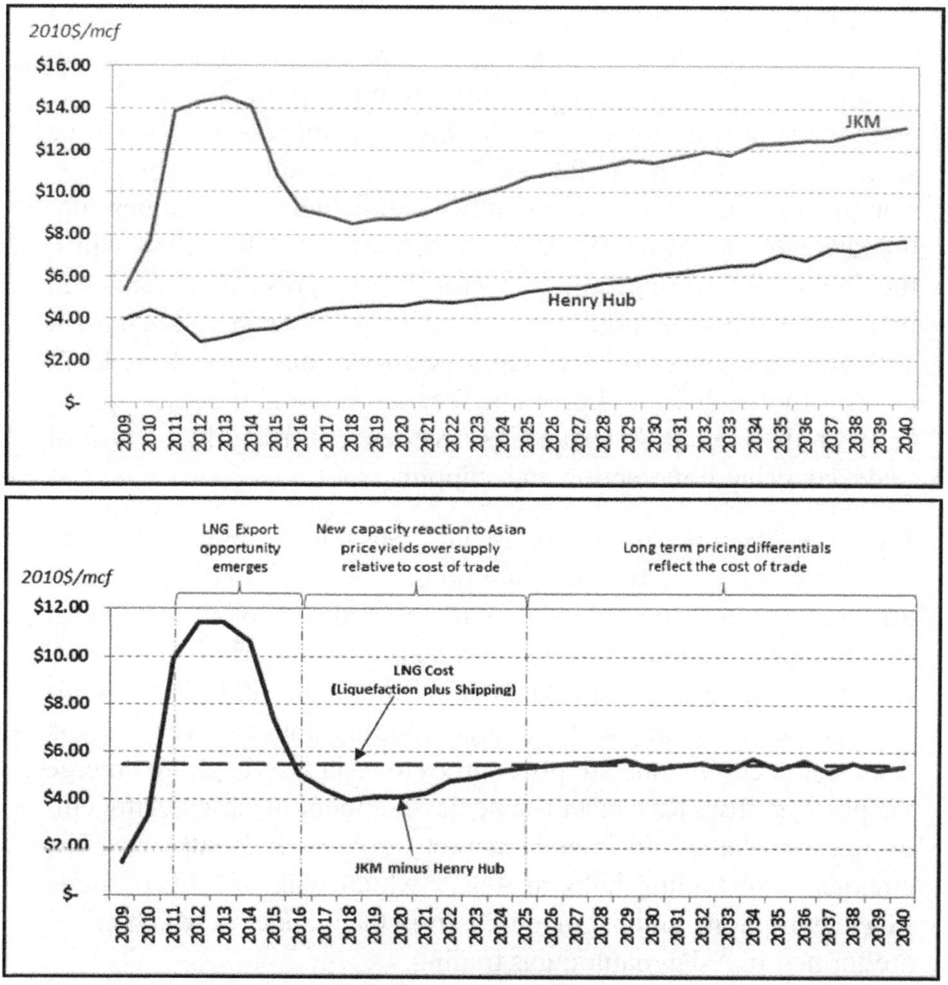

Figure 7. Long Term Projection of Prices and Spreads in Asia and the United States.[14]

Figure 7 shows that on an annual basis, the current price differential is eliminated when new LNG supplies come to market from the United States and Australia. Exacerbating matters in this exercise is the assumption of the slow reactivation of the Japanese nuclear fleet from 2015 through 2019. Importantly, however, the majority of the spot price movement is the result of new supplies being

14. RWGTM v.Feb2014, Medlock.

brought to bear, as the price impact is most dramatic between 2014 and 2017. Shale development in this case does not materialize in China in a meaningful manner until the mid-to-late 2020s. Hence, domestic supply options are not driving the compression of the price spread between Asia and the United States. In addition, while Russian pipeline supplies are ultimately developed to serve Chinese demands, they do not matriculate into service until after 2018. Thus, the only real, meaningful impetus for the compression of the Asian premium is the relaxation of the constraint to serve the Asian market with the emergence of LNG supplies. In the long term, U.S. LNG exports approach 5.5 billion cubic feet per day, and the price spread between the United States and Asia settles in at roughly the cost of trade, covering liquefaction and shipping.

There are, of course, numerous assumptions that are contained in any model simulation—such as assumptions about economic growth in different countries that translate into demand outlooks—but the results presented herein are robust to a number of differing scenarios.[15] In fact, in the results presented herein, the model indicates significant market deepening as LNG trade roughly doubles over the next decade and continental supply sources to Asia also begin to emerge via pipeline from Russia and shale development in China. Altogether, this translates into greater market liquidity and will foster the emergence of trading hubs in Asia,[16] which will, in turn, promote financial liquidity and begin to transform the manner in which prices are formed in Asian natural gas trading.

---

15. "Geopolitics of Natural Gas: Study," James A. Baker III Institute for Public Policy, February 21, 2014, available from *bakerinstitute.org/research/geopolitics-natural-gas/*.

16. Indeed, the RWGTM simulations indicate the possible emergence of a notional hub in Shanghai, with a physical presence for hub services emerging at Xian in the Southern Ordos Basin. This is facilitated by multiple sources of gas moving through the area long-term from Russia, Central Asia, and domestic sources. A connection to the coast provides a touch point for competition between these supply sources and waterborne LNG, which only serves to enhance liquidity in the region.

Concluding Remarks

The global gas market is in a period of transition that has been prompted by the emergence of shale in North America. In fact, the emergence of shale in the United States turned the LNG market upside-down in a matter of a decade, as the consensus view that the United States would be a major LNG importer transformed into one of the United States becoming an LNG exporter. This has occurred against a backdrop in the global gas market in which the spot prices of natural gas delivered in Europe, Asia, and the United States diverge dramatically from long-standing historical relationships.

The wake of the disaster at Fukushima triggered an unexpected rise in demand for waterborne LNG supplies. Constraints on the ability to meet the unexpected demand shock resulted in the spot price of Asian LNG rising to unprecedented levels relative to other global spot price markers. This phenomenon is not unprecedented, as we have seen similar circumstances in the continental North American market when extreme cold in certain regions drives up local demand in excess of what existing pipeline capacity can deliver. For example, an extreme cold weather event in the Northeast has been known to trigger the daily price in Boston (at Algonquin City Gate and Tennessee Gas Pipeline Zone 6) to jump more than $70 per mcf above the price at HH because pipeline capacity is not sufficient to meet the sudden surge in heating demand. This is often referred to as a "basis blowout." Fortunately for consumers in the affected regions, these price shocks are short-lived and subside when the cold weather event passes. Moreover, the depth of the U.S. market provides substantial liquidity through which price differences are quickly arbitraged as the supply constraint is relaxed.

Unfortunately for Asian consumers, the Asian market price will not come back down until something is done to alleviate the constraints on deliverability, such as a demand reduction or the introduction of new supplies. Such an adjustment will take time to materialize because Japanese nuclear capacity is not being reactivated quickly, and construction of new LNG capacity is plagued with long lead times. However, as new supplies come online and/or nuclear capacity in Japan is brought back up, there should be a return in the Asian LNG spot price to a level that is consistent with a globally arbitraged price. Indeed, the current high price in Asia is causing most prospective LNG developers around the world to hone their focus on being able to deliver to the Asian market.

The eventual emergence of new suppliers in the global gas market makes it paramount to consider what the increased trade in LNG will do to the nature of pricing. As the United States becomes an LNG exporter, the effects on global markets could be profound. The global gas market will deepen and become physically linked to the most liquid continental natural gas market in the world. This should, in turn, promote greater trade and ultimately alter the liquidity paradigm that has dominated the Asia-Pacific market, which will alter the manner in which natural gas is priced in Asia. Moreover, when this happens it will likely do so quickly; after all, the rise to current price levels in Asia relative to prices elsewhere happened in only six months, a fact often forgotten in the discussion about future pricing in Asia.

Even if, *ex post*, the compression of the JKM-HH price differential renders U.S. LNG liquefaction investments less profitable than forecasts suggest, the establishment of a link from U.S. supplies to foreign markets will have potentially dramatic implications. A direct link between the United States and abroad will only serve to accelerate international market liquidity, thereby lowering liquidity risk. This could, all else equal, alter the financing risk of LNG projects and lower the importance of oil-linked bilateral relationships. In any case, as the story continues to unfold, the international gas market will evolve into something dramatically different from what it is today.

Finally, the spot price in Asia will remain relatively high until there is investment in new gas delivery capability or nuclear capacity in Japan is reactivated. When new supply is added or demand is reduced, thereby alleviating the deliverability constraint, price in Asia will fall very quickly. The transition will occur more quickly as market liquidity deepens once new supplies emerge. This is reflective of the manner in which constraints affect markets. As expounded above, global supply today is highly inelastic, meaning any movement in demand or additional supply will have substantial ramifications for price. The simulations in the RWGTM indicate the near term constraints contributing to short-term price pressures will likely be alleviated in the next couple of years, thus yielding significant downward pressure on price. In the long-term, price should settle at a level reflective of the marginal cost of supplies delivered into Asia, averaging $10 to $12 per mcf for the next couple of decades. This brings up a final important point: There will not be a "single price" for gas. Rather, there will remain regional price differences reflective of transportation costs, which is indeed a long-standing characteristic of the very liquid and efficient U.S. natural gas market.

CHAPTER 5

Nuclear Power: How Necessary Is It to Meet Energy Demand in Japan and Asia

Ryoichi Komiyama

Introduction

The impact of the Fukushima nuclear accident, caused by the Great East Japan Earthquake (Figure 1), was quite huge and has since caused deep discussions in formulating energy policy. The Fukushima nuclear accident has stimulated the Japanese general public's concerns over energy and environmental issues, such as the long-term power generation mix in Japan. The reliability of nuclear safety and security has deteriorated in Japan, raising drawbacks in nuclear safety standards and the imperfections of existing nuclear safety regulation and resulting in the Act on Compensation for Nuclear Damage. The problem of widespread radioactive contamination in Japan, including the relocation of many Fukushima residents to other areas, has remained virtually unsolved and has inflicted serious damage on the economy and society in Fukushima. The accident has forced the government to radically rethink the country's energy policy as well. Enormous political and technical effort is obviously required to replace the loss of nuclear power, a major base-load technology contributing to energy security and environmental sustainability before Fukushima. Reflecting on that situation, Japan needs to reformulate its climate change policy to regain its leading position in the global environmental protection initiative, while the climate change problem is becoming increasingly recognized as a constraint on the world's sustainable growth. The Fukushima accident was a tipping point for the country's energy, environmental, and nuclear policies. Japan should develop a new energy policy and technology associated with nuclear and alternative energy sources and effectively utilize them to support Asian countries in achieving sustainable growth. This chapter is organized as follows: The first section describes the energy issue in Japan before and after Fukushima; the second section reviews the nuclear energy issue in Japan and Asia; and the final section contains some concluding remarks.

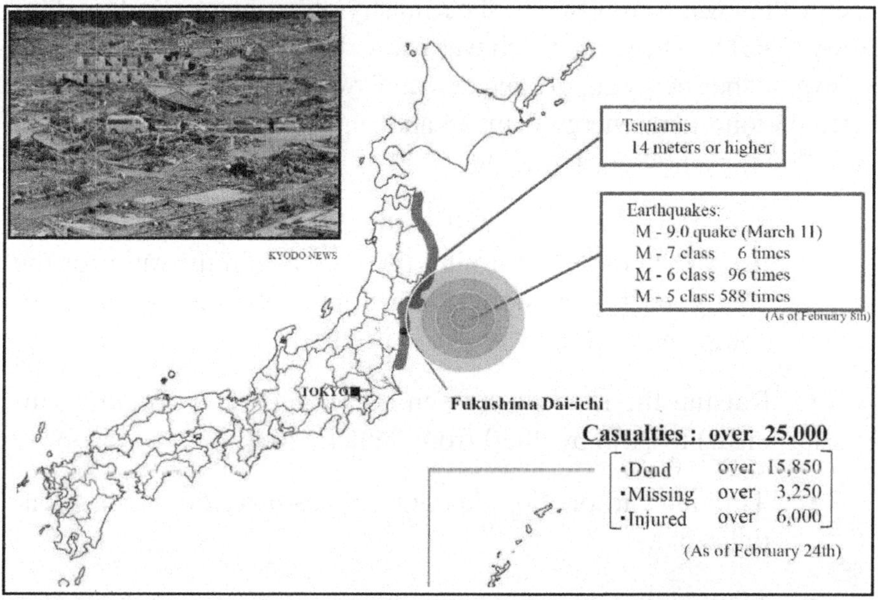

Figure 1. Great East Japan Earthquake (Enormous Earthquake, Tsunami, and Nuclear Accident): The Great East Japan Earthquake in March 11, 2011, provided an insight to the security question by demonstrating how important restructuring the energy system is.[1]

1. Ministry of Economy, Trade and Industry (METI), "Japan's Challenges Towards Recovery," March 2012, available from *www.meti.go.jp/english/earthquake/nuclear/japan-challenges/pdf/japan-challenges_full.pdf.*

Energy Issues in Japan Before and After Fukushima

National Energy Planning Before Fukushima

The basic pillars described in Japan's energy policy appear in the Basic Act on Energy Policy of 2002.[2] The Act defines the three pillars that Japanese energy policy considers: Securing a stable energy supply, assuring environmental compatibility, and utilizing market mechanisms. In order to fully achieve the provisions of the Basic Act, the Strategic Energy Plan was formulated by the Ministry of Economy, Trade and Industry (METI). The plan, which was revised in 2010, aims to effectively develop the energy supply and demand system, and it was completed based on long-term energy demand and supply outlook. It established the following ambitious targets for 2030.[3]

- Doubling the energy self-sufficiency ratio to 70% by 2030 from 38% in 2010 through the increase in the ratio for the equity energy resources developed abroad by Japanese companies, which were 26% in 2010.
- Raising the ratio of zero-emission generators (mainly nuclear) to 70% by 2030 from 34% in 2010.
- Halving carbon dioxide (CO_2) emissions from the residential sector.
- Maintaining and enhancing energy efficiency in the industrial sector at the highest level in the world. Obtaining large-scale shares of global markets for energy-related products.

---

2. Basic Act on Energy Policy, June 14, 2002, official version (in Japanese) available from *law.e-gov.go.jp/htmldata/H14/H14HO071.html.* English translation available from *faolex.fao.org/docs/pdf/jap75588.pdf.*

3. Ministry of Economy, Trade and Industry, "The Strategic Energy Plan of Japan – Meeting Global Challenges and Security Energy Futures," June 2010, summary available from *http://climateobserver.org/wp-content/uploads/2015/03/strategic-energy-plan.pdf.*

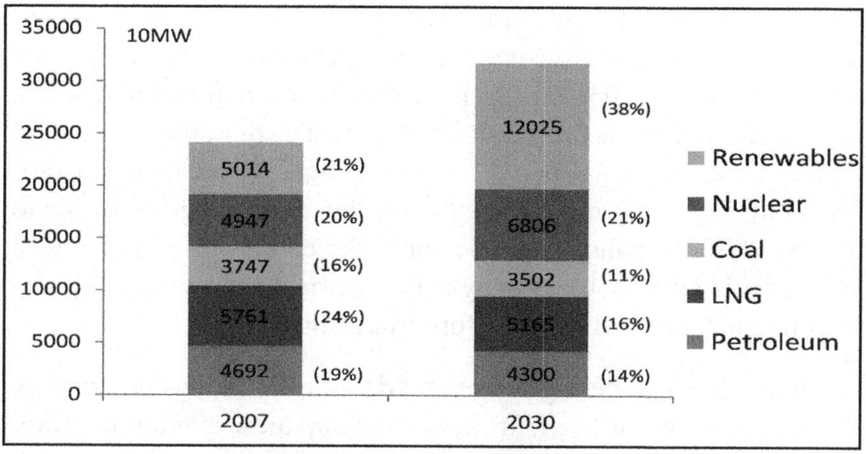

Power Generation Capacity

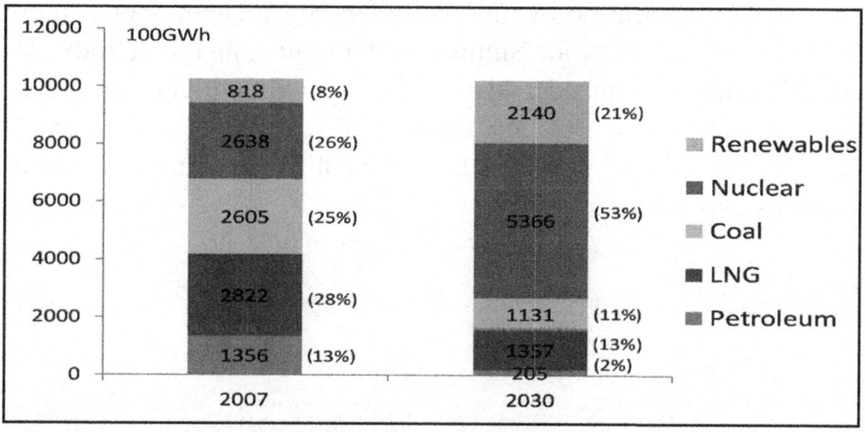

Figure 2. Targets of Power Generation Mix in the Strategic Energy Plan formulated in 2010.[4]

However, after experiencing the Fukushima nuclear accident in March 2011, the Japanese government made an important decision to review the Strategic Energy Plan.[5] This review process

4. Ibid.

5. National Policy Unit, Energy and Environment Council, "Innovative Strategy for Energy and Environment," Tokyo, September 14, 2012, Japanese version available from *www.cas.go.jp/jp/seisaku/npu/policy09/pdf/20120914/20120914_1.pdf.*

is ongoing, especially focusing on the share of nuclear energy in the country's future power generation mix. Figure 2 represents the energy outlook to 2030 in the Plan developed before Fukushima. The nuclear policy in the pre-Fukushima Plan put an emphasis on building nine additional nuclear plants to achieve a capacity factor of 85% by 2020 and more than 14, further enhancing the factor to 90%, by 2030. Expansion of nuclear is the central pillar to achieve the country's sustainable energy mix ensuring energy security and environmental compatibility before Fukushima.

In addition, the Japanese government developed a long-term energy technology policy as well. In 2007, the Japanese government announced "Cool Earth 50," a cooperative initiative to reduce worldwide greenhouse emissions by 50% from current levels by 2050 through the penetration of innovative energy technologies (Figure 3).[6] At the United Nations Summit on Climate Change in 2009, Japan officially announced that the country will mitigate its greenhouse gas emissions by 25% from 1990 levels by 2020 if an effective international framework was established where all major countries participate.[7]

6. Prime Minister of Japan Shinzo Abe, "Invitation to 'Cool Earth 50:' 3 Proposals, 3 Principles," Speech at an international conference on "The Future of Asia," Tokyo, May 24, 2007, available from *japan.kantei.go.jp/abespeech/2007/05/24speech_e.html*.

7. Prime Minister of Japan and His Cabinet, "Statement by Prime Minister Yukio Hatoyama at the United Nations Summit on Climate Change," New York, September 22, 2009, available from *japan.kantei.go.jp/hatoyama/statement/200909/ehat_0922_e.html*.

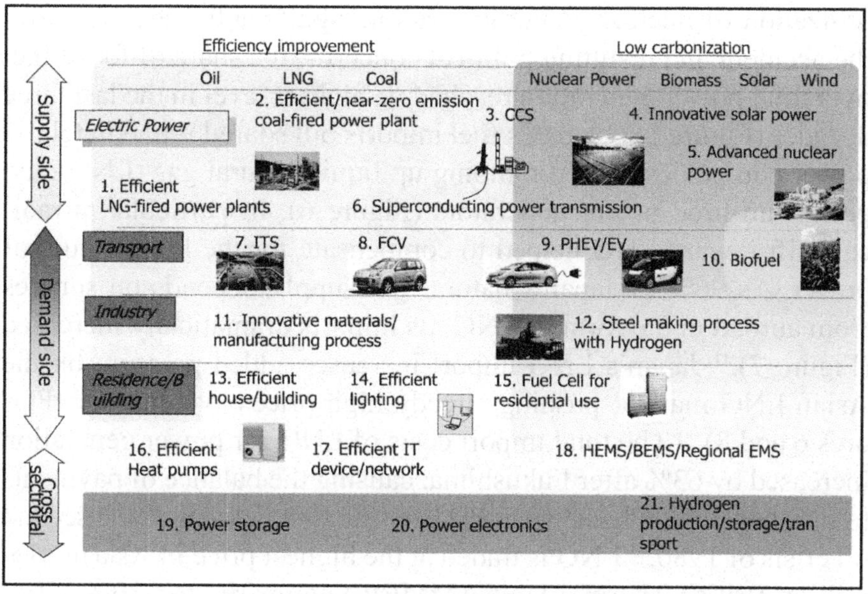

Figure 3. Key Innovative Technologies in Japan.[8]

Energy Issues in Japan after Fukushima

The Fukushima nuclear accident and the eventual difficulty in promoting the use of nuclear in the energy mix have forced Japan to face many challenges. Japan should tackle more than just overcoming those challenges and be required to develop a new prototype of energy and environmental policy that will generate sustainable growth internationally.

Fossil Fuels. The Fukushima nuclear accident caused the shutdown of all the country's nuclear power plants, which accounted for 30% of the country's electricity supply at the time (Figure 4).[9] Since the

8. Ministry of Economy, Trade and Industry, "Cool Earth – Innovative Energy Technology Program," March 2008, available from *www.meti.go.jp/english/newtopics/data/pdf/031320CoolEarth.pdf*.

9. Federation of Electric Power Companies of Japan, "Graphical Flip-chart of Nuclear Energy Related Topics 2014," 2014, pp. 1, 2, 7, available from *www.fepc.or.jp/english/library/graphical_flip-chart/__icsFiles/afieldfile/2014/07/07/*

utilization of nuclear power generation significantly declined after the accident and resulting political concerns, the share of fossil fuel over total power generation reached its highest level in the last three decades (Figure 5).[10] Japan's fuel imports bill soared immediately as power companies started ramping up liquid natural gas (LNG) and petroleum-fired power generators (Figure 6). In particular, a radical shift toward LNG helped to compensate for the loss of nuclear power. As 96% of Japan's natural gas supply depends on sources from abroad in the form of LNG, its imports dramatically increased (Figure 7).[11] Japan's LNG import increases added pressure on the Asian LNG market, pushing already high prices even higher (Figures 6 and 8).[12] The total import costs of LNG for power generation increased by 63% after Fukushima, causing the balance of payments to turn negative in fiscal year 2011 for the first time since the second oil crisis of 1980.[13] LNG is traded at the highest price in Asia at $15/million British Thermal Units (MMBtu) compared to $10/MMBtu in Europe, while U.S. natural gas is priced at around $5/MMBtu.

---

all_english_2014.pdf.

10. Ministry of Economy, Trade and Industry, "Monthly Power Generation Statistics," 2010-2014.

11. Ibid; Ministry of Finance, "Monthly Trade Report," 2009-2014; and Ministry of Economy, Trade and Industry, "Monthly Resource and Energy Report," 2009-2014.

12. Institute of Energy Economics Japan, *EDMC Handbook of Energy & Economic Statistics in Japan*, 2013; Kazuhiko Ogimoto, Izumi Kaizuka, Yuzuru Ueda, and Takashi Oozeki, "A Good Fit: Japan's Solar Power Program and Prospects for the New Power System," *IEEE Power and Energy Magazine* 11, no. 2, March/April 2013, pp. 65-74, available from *web.mit.edu/12.000/www/m2018/pdfs/japan/solar.pdf; and TradingCharts.com*, "2011 Natural Gas (Globex) Historical Prices/Charts (NYMEX)," available from *futures.tradingcharts.com/historical/NG_/2011/0/continuous.html.*

13. Institute of Energy Economics Japan, *EDMC Handbook*, 2013.

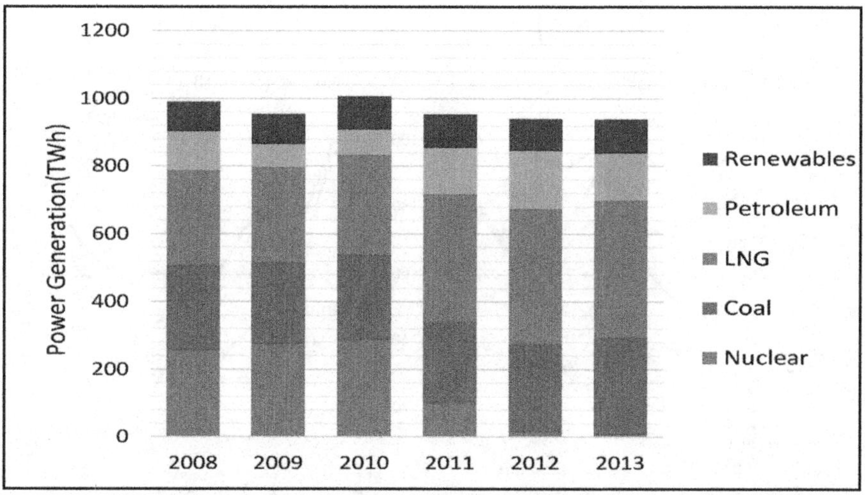

Figure 4. Power Generation Mix: Nuclear is replaced by LNG and petroleum after Fukushima.[14]

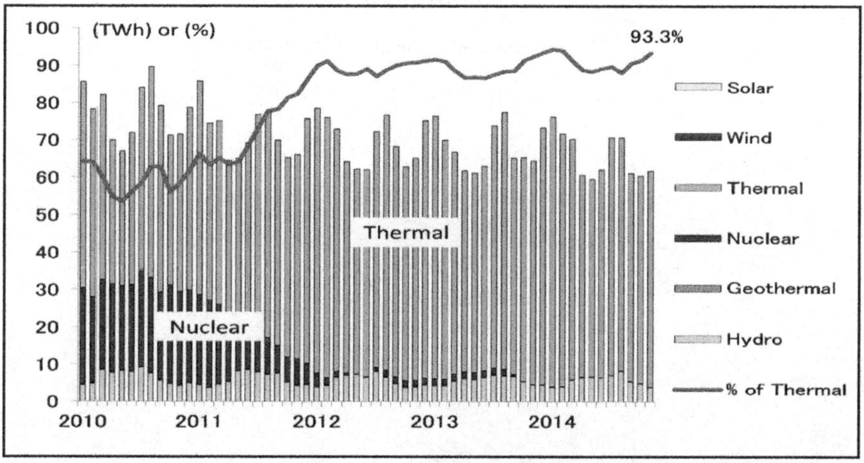

Figure 5. Monthly Power Generation Mix: Since the utilization of nuclear power generation significantly declined after the earthquake due to the accident and political concerns, the share of fossil fuel over total power generation reached the highest level in the last three decades.[15]

---

14. Federation of Electric Power Companies of Japan statistics, available from *www.fepc.or.jp/about_us/pr/pdf/kaiken_s3_20160520_1.pdf*

15. Ministry of Economy, Trade and Industry, Monthly Power Generation Sta-

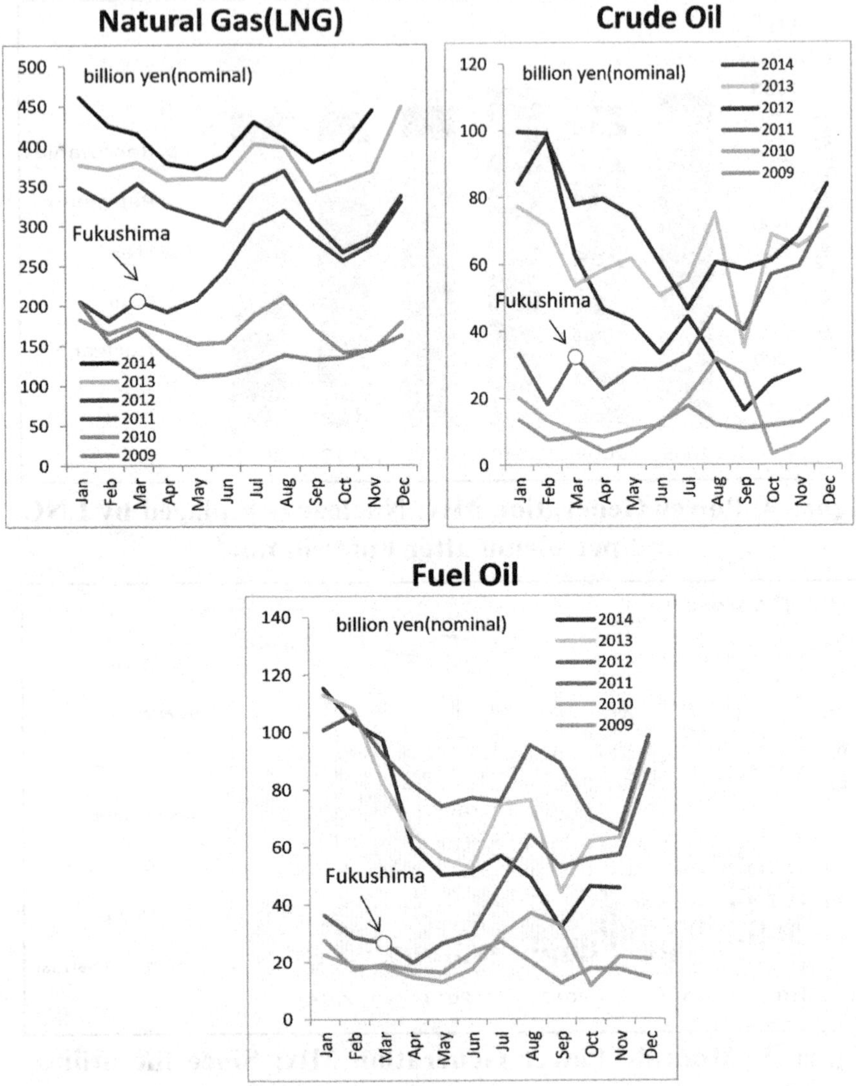

Figure 6. Fuel Cost for Power Generation in Japan: Nuclear suspension after Fukushima caused soaring fuel-import costs for power generation.[16]

---

tistics, 2014.

16. Ministry of Economy, Trade and Industry, Monthly Electricity Report, 2014; MOF, Monthly Trade Report, 2014; METI, Monthly Resource and Energy Re-

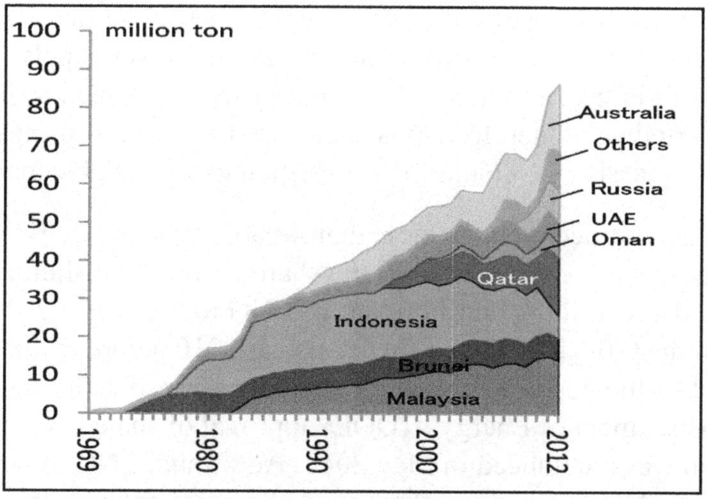

Japan's LNG imports.

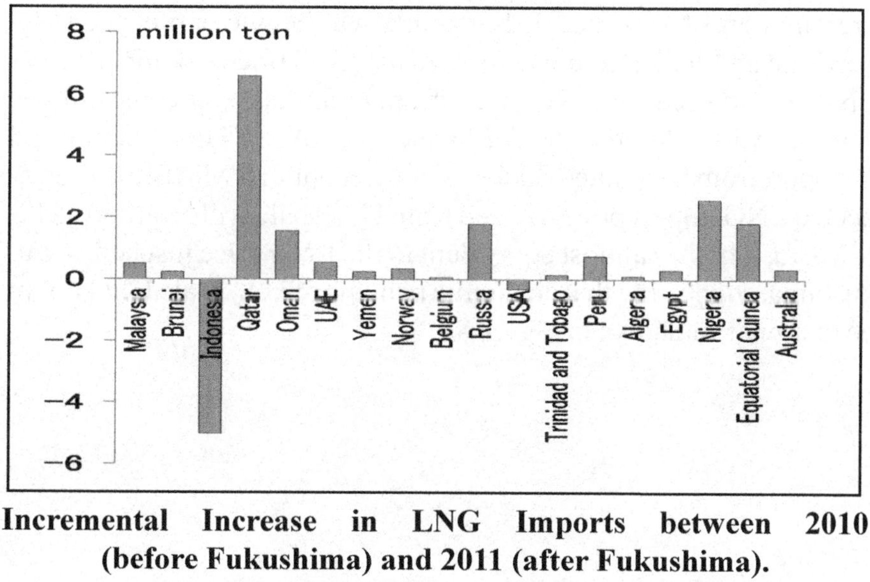

Incremental Increase in LNG Imports between 2010 (before Fukushima) and 2011 (after Fukushima).

Figure 7. Rising LNG Imports before and after the Fukushima Nuclear Disaster.[17]

port, 2014.

17. Ministry of Finance, *Monthly Trade Report*, 2013 and METI, Monthly *Report of Resource and Energy*, 2013.

LNG is the most important fuel to supply electricity after the Fukushima accident. To compensate for the gap in power supply, Japan, the world's largest importer of LNG, has to depend even more heavily on imported gas. Therefore, it is necessary to take various measures to assure massive LNG imports in an efficient and reliable manner.

LNG currently accounts for more than 40% of Japan's electricity supply, a nearly 30% increase from the share before Fukushima. Japan imported a record 87.3 million tons of LNG in fiscal year 2012 after Fukushima (70.0 million tons in fiscal year 2010 before Fukushima), as most of the country's nuclear reactors remained shut down. The U.S. Department of Energy's (DOE) approval of shale LNG exports to Japan was announced in May 2013. Additional LNG imports derived from North American shale gas will help to enhance the energy security of Japan. However, Japan is far from resolving the Asian premium on LNG prices. U.S. exports will be one part of the LNG portfolio and Japan should recognize the LNG price risk for ensuring LNG import security. LNG prices from U.S. shale sources should be cautiously monitored, considering the costs of LNG liquefaction and transport from the United States to Japan (about $6/MMBtu). The expected LNG import price derived from U.S. shale will be around $12/MMBtu, which is almost equivalent to the LNG price just before Fukushima, suggesting that economic benefits of U.S. shale LNG might be marginal, as shown in Figure 8.[18]

18. Institute of Energy Economics Japan, *EDMC Handbook*, 2013; Chicago Mercantile Exchange, "Commodity Futures;" U.S. Energy Information Administration, "Natural Gas Spot and Futures Prices (NYMEX)," available from *www.eia.gov/dnav/ng/ng_pri_fut_s1_d.htm*; and U.S. Energy Information Administration, Short-term Energy Outlook (STEO), available from *www.eia.gov/forecasts/steo/outlook.cfm*.

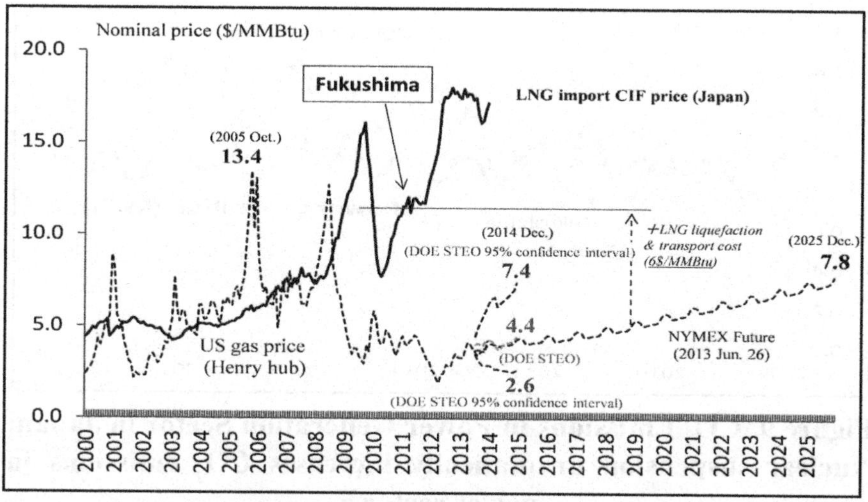

Figure 8. LNG import Price in Japan and U.S. Gas Price (Henry Hub)[19]

CO_2 Emissions. After the Fukushima accident, heavy reliance on fossil fuels has caused rapid increases in Japan's CO_2 emissions. The utility power companies accounted for 439 million tons of CO_2 for the year 2012 (after Fukushima),[20] up 17% from 374 million tons in the year 2011 (before Fukushima, see Figure 9). Kansai Electric Power Company, which relies most on nuclear power, produced 65.7 million tons, a "40 percent increase in CO_2 emissions."[21] Japan must install less carbon-intensive alternative energy to replace nuclear power. Otherwise, the increased use of fossil fuels, especially LNG and petroleum, will make it harder for Japan to mitigate the country's CO_2 emissions and meet its target of a 25% reduction in greenhouse gasses by the end of the decade.

---

19. Henry hub price: EIA/DOE, DOE short-term outlook: EIA/DOE STEO, NYMEX future price: NYMEX Market Data, LNG price (Japan): Institute of Energy Economics Japan (IEEJ).

20. Institute of Energy Economics Japan, *EDMC Handbook*, 2013.

21. Kansai Electric Power Company, "CSR Report 2012", Japanese version available from *www.kepco.co.jp/corporate/list/report/report2012.html*.

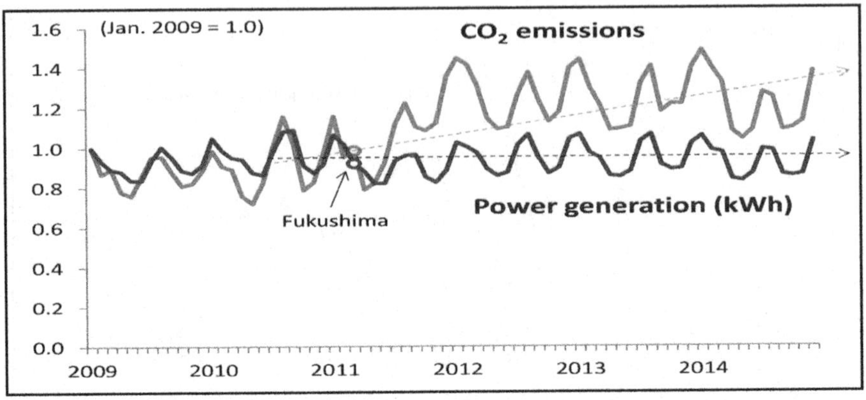

Figure 9. CO_2 Emissions in Power Generation Sector in Japan: Nuclear suspension dramatically increases CO_2 emissions in power sector.[22]

Renewables. The feed-in tariff (FIT) introduced in July 2012 is working as a powerful driver to increase renewable energy supplies after Fukushima. The rate of its recent development goes beyond expectations. In August 2011, the Act on Purchase of Renewable Energy-Sourced Electricity by Electric Utilities was passed by the Japanese Parliament. This Act took effect on July 1, 2012, and orders electric utilities to purchase electricity generated from renewable energy sources (solar photovoltaics [PV], wind power, small and medium-sized hydropower, geothermal, and biomass) based on fixed term contracts with fixed prices from 10 to 20 years. Costs incurred by the utilities in purchasing renewable energy-sourced electricity are transferred to all electricity customers, who pay a surcharge for renewable energy at a rate proportional to their electricity usage. Contract rates and periods are set by the Ministry of Economy, Trade and Industry.

FIT has provided outstanding progress in solar PV installation. The accumulated installed capacity of photovoltaic systems was only 4.9 GW in 2011.[23] The capacity jumped by 2.0 GW in the following year (Figure

22. Estimated from METI, "Monthly Electricity Report," 2014.

23. Institute of Energy Economics Japan, *EDMC Handbook*, 2013.

10). Additionally, more than 10 GW of PV capacity has been approved and is waiting for actual installation. In 2015, PV capacity of 12 GW was newly installed. Based on this estimate, Japan will outstrip Germany in the annual installation of photovoltaic systems. This boom is noticeable only in photovoltaic power as it is easily installable compared with other renewables, and the level of FIT is significantly higher. It was set at ¥42(42¢)/kWh for the first year (FY2012), and ¥38(38¢)/kWh for the second year (FY2013). However, the progress in geothermal and wind power is rather lower due to the shortage of sufficient grid capacity and appropriate locations. Particularly, geothermal faces a serious problem as most of the resources (about 80%) are located within natural areas, where environmental regulations will not fully allow geothermal development.

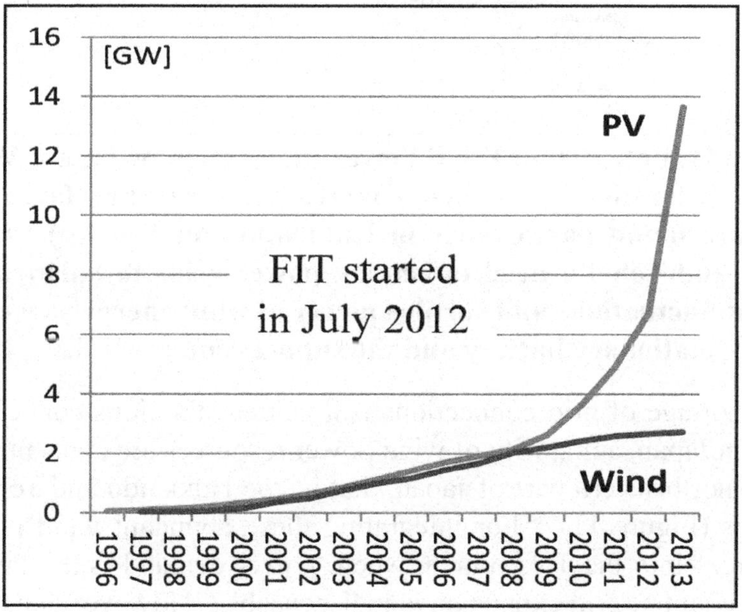

Figure 10. PV & Wind Capacity in Japan.[24]

24. Institute of Energy Economics Japan, *EDMC Handbook of Energy & Economic Statistics in Japan*, 2014.

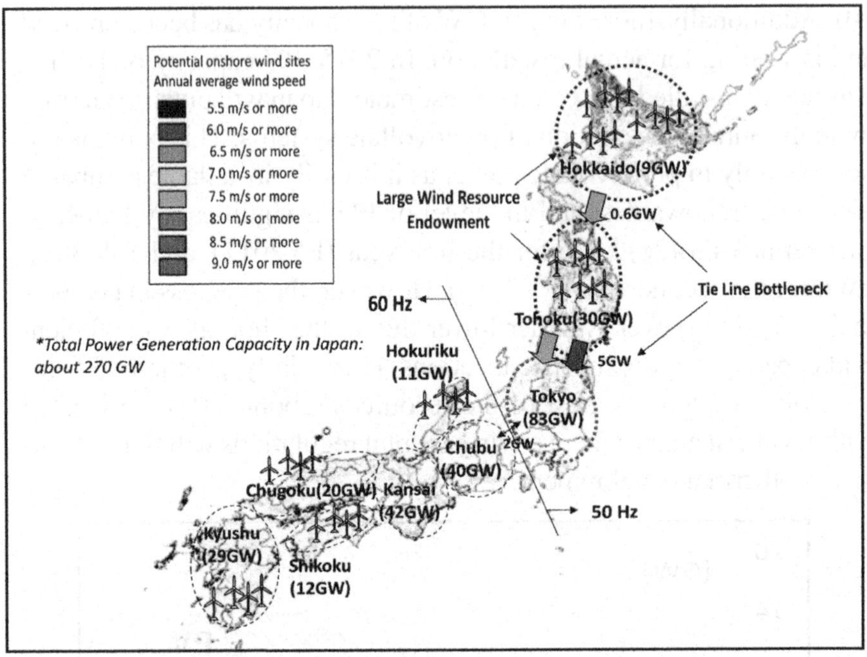

Figure 11. Potential of Wind Power Integration in Japan: Wind power generation is expected to increase in the future (favorable sites are found particularly in Hokkaido and Tohoku). Hokkaido and Tohoku need to employ power grids to balance the output fluctuations and surplus power of wind energy as well as stationary battery and the suppression control.[25]

The shortage of grid connections is a matter of serious concern as well. In Japan, a majority of wind power resources are concentrated in the northeastern part of Japan, that is, the Hokkaido and Tohoku regions (Figure 11).[26] For integrating these abundant wind power resources into the Tokyo area, where power demand is the largest in the country, grid expansion is indispensable. However, Japanese power companies are allowed to have a regional monopoly with ver-

25. Ministry of Environment Japan, "Study of Potential for the introduction of Renewable Energy," March 2010, available from *www.env.go.jp/earth/report/h22-02/full.pdf*; English summary available from *www.env.go.jp/earth/report/h22-02/09-summary.pdf*.

26. Ibid.

tically integrated systems. Therefore, the grid system has been centrally planned, allowing very limited opportunity for additional renewables. Furthermore, the financial situation of power companies is not good enough to expand grid capacity due to the shutdown of most nuclear power plants and the corresponding limited financial resources.

Massive deployment of intermittent sources, such as wind and PV (Figures 12 and 13),[27] to replace nuclear is another hard challenge for grid management. According to the power system analysis under large-scale PV and wind penetration, it is observed that renewable variability is technically controllable by power charge and discharge of rechargeable battery and pumped-hydro storage flexible load following operation by thermal power plants, and the output suppression control of PV and wind, which implies that various measures dynamically serve as a whole to regulate the short-cycle variation of the renewables power generation (Figure 14).[28] Thus, the dynamic operation of these supply-side technologies is crucial in increasing the ratio of intermittent sources that can be integrated in the power grid. However, these types of operations are totally different from the current one in which power demand and supply are balanced through base-load, middle-load, and peak-load generators in the day-ahead electricity market in a stable way, and the technical feasibility of these operations should be cautiously investigated.

---

27. Ryoichi Komiyama and Yasumasa Fujii, "Assessment of massive integration of photovoltaic system considering rechargeable battery in Japan with high time-resolution optimal power generation mix model," *Energy Policy* 66, March 2014, pp. 73-89.

28. Ryoichi Komiyama and Yasumasa Fujii, "Assessment of Japan's Optimal Power Generation Mix Considering Massive Deployment of Variable Renewable Power Generation," *Electrical Engineering in Japan* 185, no. 2, November 2013, pp. 1–11 and Ryoichi Komiyama, Saeko Shibata and Yasumasa Fujii, "Assessment of Optimal Power Generation Mix Considering Extensive Variable Renewable Energy and Nationwide Power Interchange through Tie Lines," *Electrical Engineering in Japan* 188:, no. 3, August 2014, pp. 11–21.

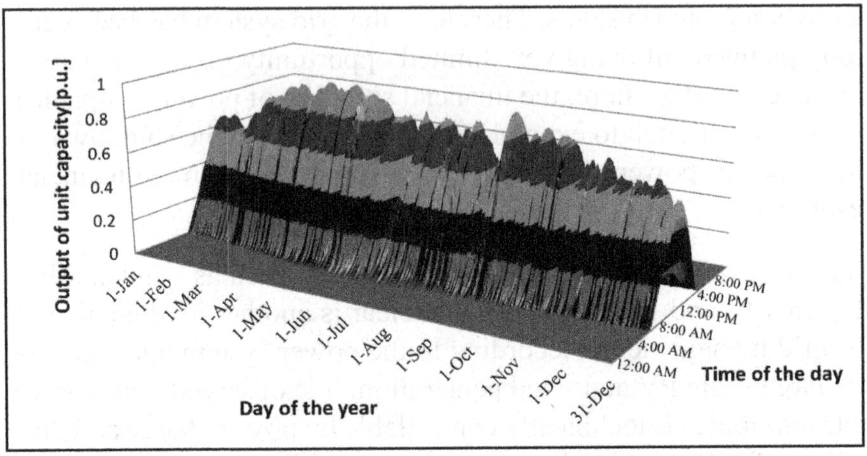

Figure 12. PV Output Pattern of Japan: Horizontal axis represents day of the year (365 days), and normal axis, time of the day in 10-minute time resolution.

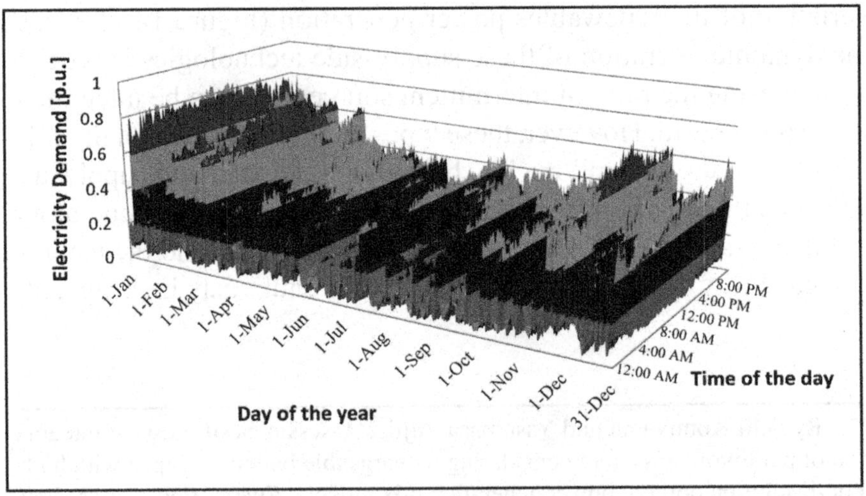

Figure 13. Wind Output Pattern of Japan: Horizontal axis represents day of the year (365 days), and normal axis, time of the day in 10-minute time resolution.

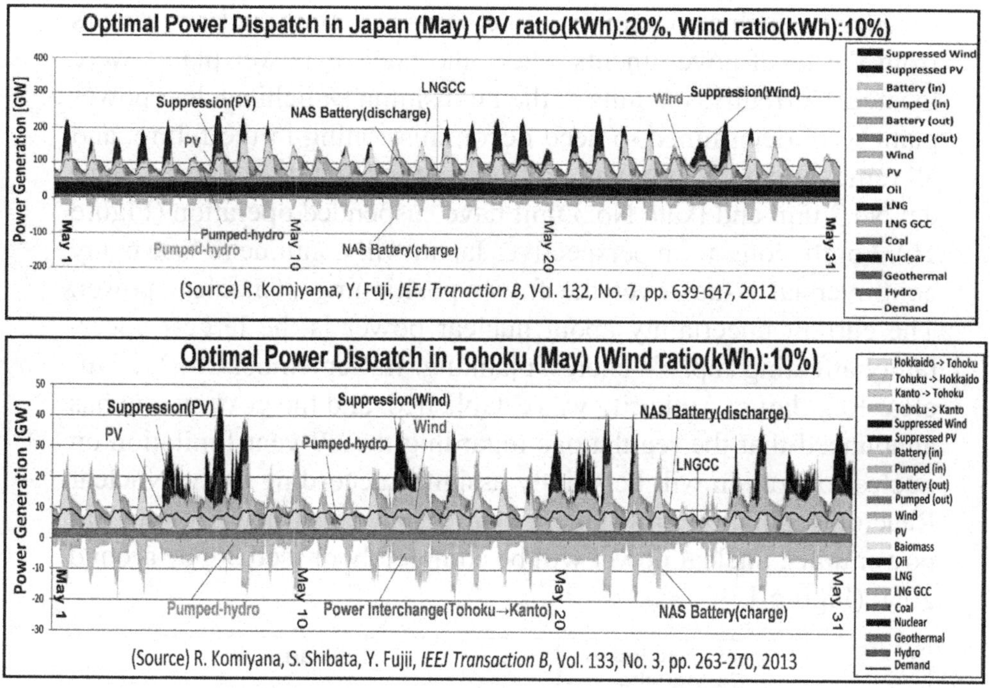

Figure 14. Large-scale Integration of Variable Renewables in Japan (Simulation Analysis).

Nuclear Energy in Japan and Asia

Nuclear Energy in Japan

Japan's nuclear energy policy is currently under review after the Fukushima nuclear accident. In the accident, four nuclear power plants located on the country's Pacific coast were severely damaged by a tsunami, which caused a loss of electric power and cooling capability (Figure 15). In the months after the Fukushima accident, each of the nuclear units was incrementally shut down for scheduled inspection, which is mandated once every 13 months. After the shutdowns for regulatory inspection, the utility companies did not restart operations due to the difficulty in getting public acceptance such as obtaining agreements to restart nuclear reactors from the local governments where the nuclear power plants were located. Currently, six units of the Fukushima Daiichi nuclear power plant were determined to need decommissioning in the future, and all remaining nuclear power units except for Sendai No.1 unit, Sendai No.2 unit and Ikata No.3 unit have suspended operation (Figure 16). For the long-term perspective, Japan will continue to face many controversial issues over the future positioning of nuclear power. The current uncertainty about nuclear power is the largest uncertainty affecting Japan's energy planning. In September 2012, a Nuclear Regulation Authority was established, and the government has announced that the regulations regarding the 40-year limitation on nuclear operation will be strictly assigned according to the Nuclear Reactor Regulation Law; unless there is new construction of nuclear power units, nuclear power will be completely phased out by around 2050 (Figure 17).

	Great East Japan Earthquake			Tsunami				
	Electric Power Supply		Cooling system	Electric Power Supply		Cooling system	Height of Tsunami (m)	Height of the Site (m)
	From off-site	Backup generators	Seawater pumps	From off-site	Backup generators	Seawater pumps		
Onagawa #1,#2,#3	○	○	○	○	○	○	13	13.8
Fukushima Daiichi #1,#2,#3, #4,#5,#6	×	○	○	Earthquake ×	Unit #1 - #5 × Unit #6 ○	Unit #1 - #4 × Unit #5,# 6 ×→○ (after few days)	15.5	10 (Unit#1- #4) 13 (Unit #5, #6)
Fukushima Daini #1,#2,#3,#4	○	○	○	○	Unit #1, #2 × Unit #3, #4 ○	Unit #1,#2,#4 ×→○ (after few days) Unit #3 ○	14.5	12
Tokai Daini #1	×→○ (after few days)	○	○	Earthquake ×→○ (after few days)	○	○	5.3	8

Figure 15. Overview of Fukushima Daiichi NPP Accident: Four nuclear power plants are located on the coast facing the Pacific Ocean, but the loss of electric power and cooling capability happened only at the Fukushima Daiichi Nuclear Power Station. Nuclear power systems remained intact in the other three nuclear power stations.

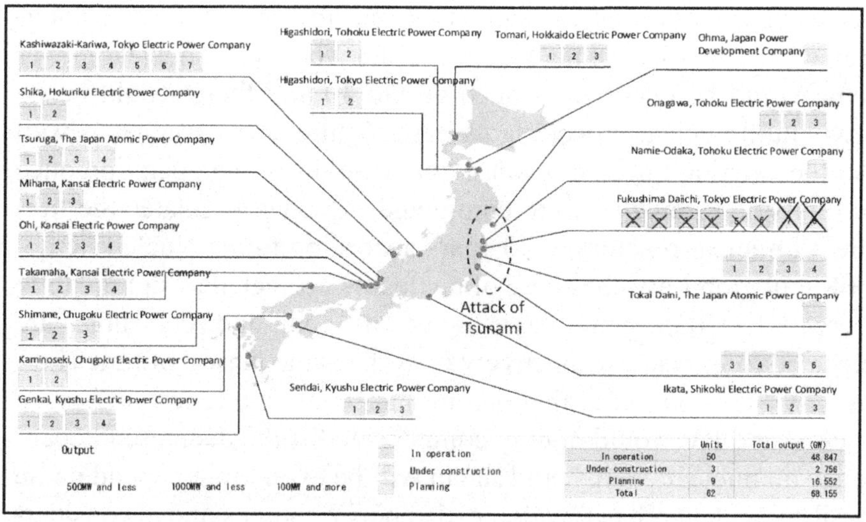

Figure 16. Nuclear Power Plants in Japan after Fukushima: Nuclear has been a very important power source in Japan, now shut down one by one for regulatory inspection.[29]

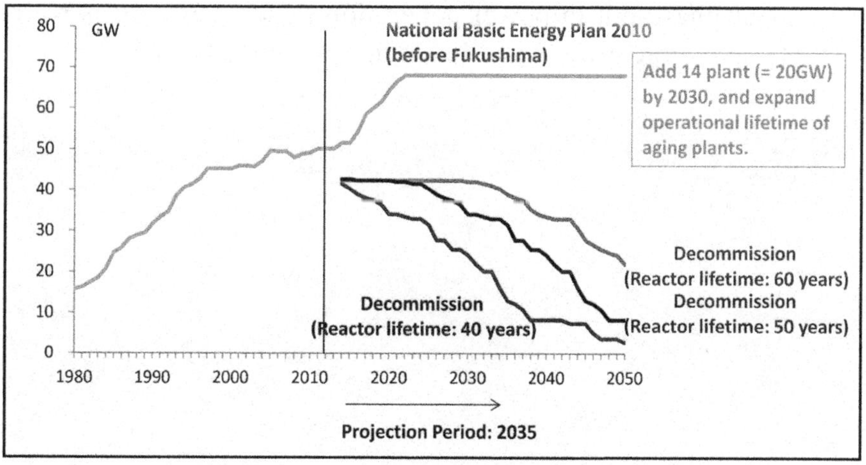

Figure 17. Nuclear Scenarios in Japan: Japanese nuclear regulation authority declared that the operational period of nuclear plants is legally limited to 40 years with the possibility of a less-than-20-year extension.

29. FEPC (Federation of Electric Power Companies Japan) and IEEJ.

Hence, it is important to understand the impact on Japan's long-term energy mix by the nuclear phase-out where each nuclear power plant would be decommissioned at age 40 and there would be no newly built nuclear power plants in the future. It is assumed as well that the existing nuclear power plants would resume operation once current safety reviews were completed. If Japan regulates the lifetime of nuclear reactors to 40 years according to the Nuclear Reactor Regulation Law, the share of nuclear is projected to significantly decrease by 2035, while the share of gas, coal, and renewables are expected to increase respectively over the same period of time (Figure 18).[30] It is observed that natural gas, coal, and renewables such as wind and PV would mainly compensate for the decreased dependency on nuclear energy, and Japan will be even more dependent on fossil fuels. Japan has few domestic energy sources and is considerably dependent on imported fuel (Figure 18). Ensuring firm access to secure the equity of energy resources becomes an increasingly important challenge for Japan under the nuclear phase-out policy. The nuclear phase-out in power generation raises concerns as well for stable economic growth and its environmental sustainability.

30. Asia Pacific Energy Research Centre, Institute of Energy Economics, Japan, *APEC Energy Demand and Supply Outlook*, Vol. 2, 5th Edition, Singapore: Asia-Pacific Economic Cooperation, February 2013, pp. 83-84, available from *publications.apec.org/publication-detail.php?pub_id=1389.*

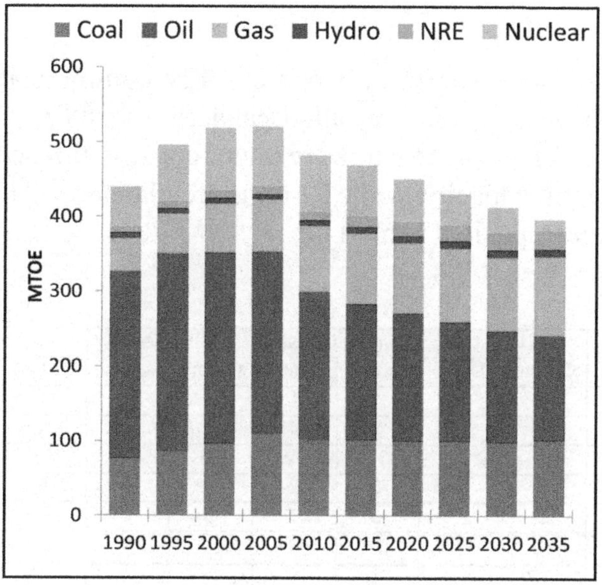

Primary Energy

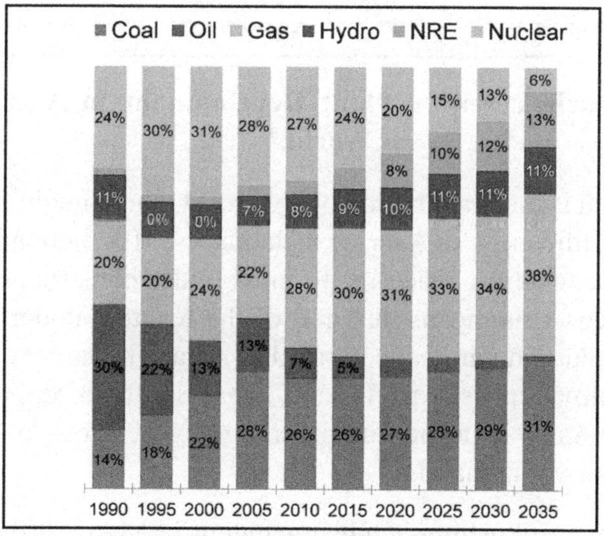

Power Generation Mix

Figure 18. Japan's Energy Outlook to 2035 Under Gradual Phase Out of Nuclear Power Plant.[31]

31. Ibid.

Nuclear Energy in Asia and the World

Around the world, 31 countries possess 429 commercial nuclear power reactors with a total installed capacity of 388 GW, and 76 additional nuclear power reactors are under construction, equivalent to 20% of present capacity, while 97 units are planned, equal to one-third of present capacity (Table 1).[32]

Output [GW]

	In Operation		Under Const.		Planned		Total	
	Output	Units	Output	Units	Output	Units	Output	Units
Japan	46	50	4	4	12	9	63	63
South Korea	21	23	5	4	7	5	33	32
China	13	15	35	32	26	23	73	70
Chinese Taipei	5	6	3	2	0	0	8	8
India	5	20	5	7	5	4	15	31
Asia Total	89	114	53	49	51	41	193	204
USA	107	104	1	1	11	9	118	114
France	66	58	2	1	0	0	68	59
Russia	25	29	10	11	18	17	54	57
World Total	388	429	78	76	111	97	577	602

Table 1. Nuclear Power Plant Development in Asia and the World.[33]

Notably, Asia is the main region of the world where nuclear power is growing significantly. In Asia, as of January 2016, there are 124 operating nuclear power reactors, 40 units under construction and 44 units planned. This means that half of the reactors under construction and in planning are concentrated in Asia. The largest growth in nuclear deployment is expected in China, South Korea, and India. Many other Asian countries are planning new reactors to meet their increasing power demands.

Even after the Fukushima nuclear accident in March 2011, consid-

32. Japan Atomic Industrial Forum, "World Nuclear Power Plants," May 21, 2013, English summary available from *www.jaif.or.jp/english/news_images/pdf/ENGNEWS02_1369269416P.pdf*.

33. Ibid.

erable growth of nuclear energy utilization on the part of Asian countries is projected in the long-term perspective (Figure 19).[34] This increase is attributable to the economic and environmental advantages of nuclear energy as well as stringent nuclear safety regulation intensified after Fukushima. The economic advantages of nuclear power consist of its affordable price and lower risk of fuel price fluctuations compared to fossil fuels. The environmental merits include its low carbon intensity in the power generation process. The principal obstacle to nuclear promotion is lower public acceptance associated with safety, security, and safeguard (3S) issues. The Fukushima accident has caused increased concerns about nuclear safety, focusing on the resilience of nuclear facilities during a large natural disaster. Consequently, an enormous technical and political effort will be necessary to resolve these concerns and recover public confidence in the safety and security of nuclear reactors to supply electricity in a more sustainable way; nuclear policy should place even more emphasis on developing advanced nuclear technologies including front-end and back-end technologies and upgrading nuclear safety standards after Fukushima.

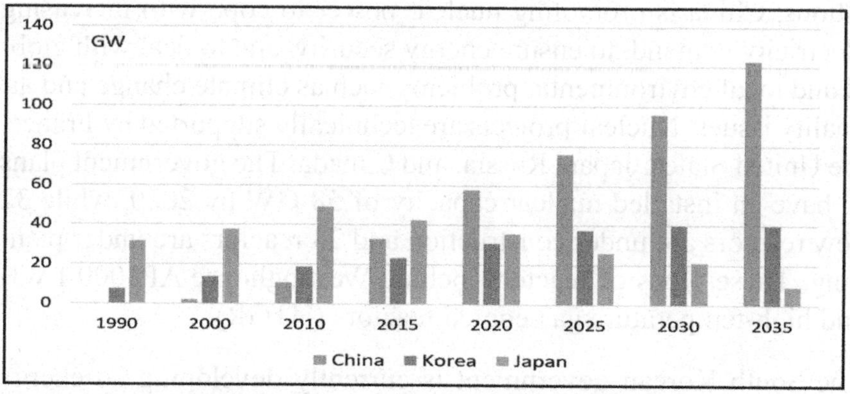

Figure 19. Outlook of Nuclear Capacity in East Asia.[35]

34. *APEC Energy Demand and Supply Outlook*, Vol. 1, p. 138.

35. Ibid.

Nuclear Energy Outlook in East Asia

The nuclear expansion in Asia is expected to be mainly concentrated in China and South Korea (Figure 19). Both countries have adopted policies encouraging progressive promotion of nuclear power. Figure 19 shows projected nuclear capacity by East Asian countries. China is expected to show the largest growth in nuclear power capacity, adding 114 GW of capacity by 2035, and South Korea is projected to add 21 GW of new capacity by 2035.

For balancing increased demand and constrained supply, China has placed energy security as the top priority in its energy and environmental policy formulation. The 12th Five-Year Plan for National Economic and Social Development (2011–2015) created a program for energy security improvement with an elaborate focus on clean energy and energy efficiency.[36] The plan described targets for the future development of nuclear energy. The number of nuclear power plants would increase from the current 15 to 25 by 2015. After Fukushima, China immediately suspended approval of new construction, implemented a nuclear safety inspection of existing nuclear reactors, and introduced more stringent safety standards and regulations. China is promoting nuclear power to cope with increasing electricity demand, to ensure energy security, and to deal with global and local environmental problems such as climate change and air quality issues. Nuclear projects are technically supported by France, the United States, Japan, Russia, and Canada. The government plans to have an installed nuclear capacity of 58 GW by 2020, while 32 new reactors are under construction and 23 reactors are under planning. These types of reactors include Westinghouse AP1000 LWR and high-temperature gas-cooled reactors (HTGR).

The South Korean government is currently developing an energy policy to encourage sustainable development. Among the goals, energy security will remain a critical issue for South Korea, since

36. Asia Pacific Energy Research Centre, Institute of Energy Economics, Japan, *APEC Energy Overview 2012*, Singapore: APEC, 2013, available from *aperc.ieej.or.jp/file/2013/6/28/APEC_Energy_Overview_2012.pdf*.

it depends heavily on fuel imports like Japan. Therefore, nuclear power is regarded as an important option for reducing fuel import dependency. South Korea has been continuously expanding nuclear power for energy security, in line with its industrial policy to export nuclear technology to achieve economic growth. South Korea has plans to build 12 new nuclear reactors, and the national plan is to expand to 35 nuclear power reactors by 2030. However, the nuclear expansion may not progress as planned due to difficulties in public acceptance; the acceptance of new reactor construction is still a significant obstacle in local communities. Collaborating with U.S. companies, South Korea developed the OPR-1000 nuclear reactor, and an AP1400 plant will be exported to the United Arab Emirates in a $20 billion deal after tough competition against France and Japan.[37]

37. World Nuclear Association, "Asia's Nuclear Energy Growth," October 2013, available from *www.world-nuclear.org/info/Country-Profiles/Others/Asia-s-Nuclear-Energy-Growth/*.

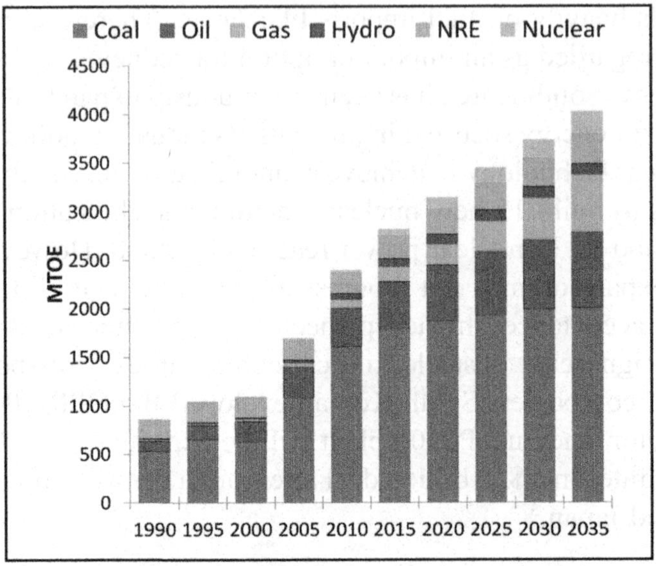

Primary Energy.

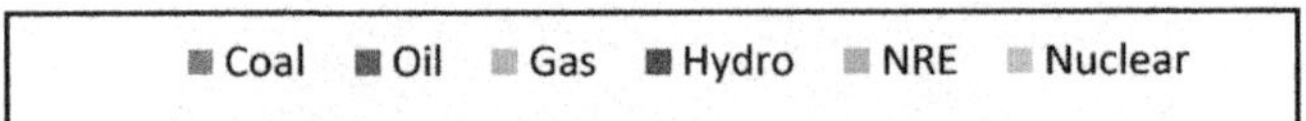

Power Generation Mix.

Figure 20. Chinese Energy Outlook to 2035.[38]

38. *APEC Energy Demand and Supply Outlook*, Vol. 2, pp. 43-53.

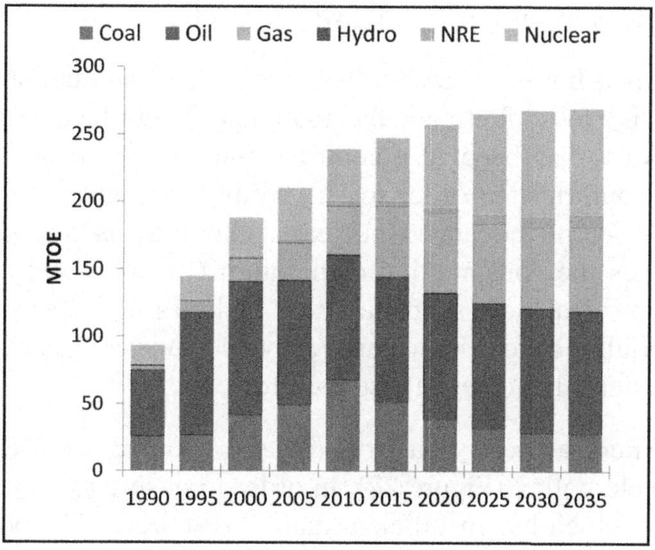

Primary Energy

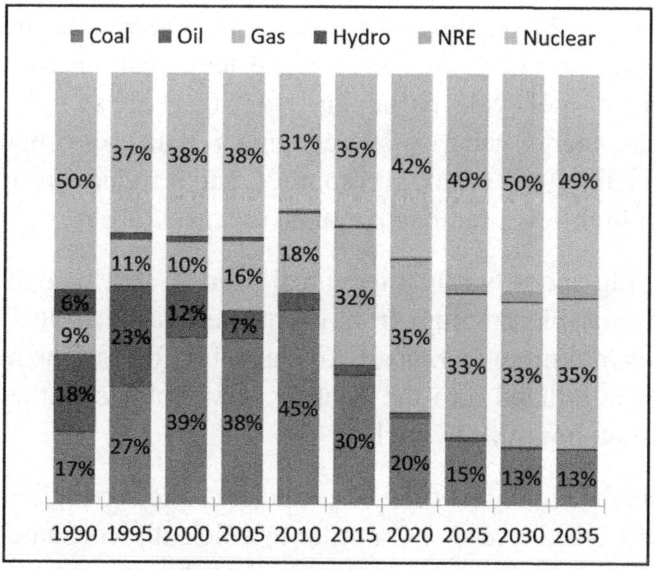

Power Generation Mix

Figure 21. South Korean Energy Outlook to 2035.[39]

39. Ibid, pp. 89-96.

Sustainable Nuclear Energy Pathway to 2100

Uranium is a finite resource whose reserves-to-production ratio is 60 years, like fossil fuels, and the treatment of high-level radioactive nuclear waste has become a controversial issue in many nuclear-installed countries. In order to sustainably continue its operation in the long-term, advanced nuclear technology is indispensable, which saves the consumption of uranium resources and mitigates the amount of nuclear spent fuel that includes high-level radioactive material; the closed nuclear fuel cycle is one of the key nuclear energy systems to achieve those requirements.

Japanese nuclear policy has traditionally adopted a closed nuclear fuel cycle policy (Figure 22), in order to ensure energy security for Japan, which has insufficient natural resources. The policy has been developed to maximize the utilization of imported uranium by domestically recycling the unburned uranium and plutonium as mixed-oxide fuel (MOX) in "pluthermal" reactors and in fast breeder reactors (FBR). Japan seeks to enhance the role of nuclear as a semi-domestic energy resource. The technical advantages of a closed nuclear fuel cycle, particularly for Japan, are straightforward: It enhances energy security by reducing reliance on imported fuels, it efficiently uses uranium resources, and it reduces a significant amount of high-level radioactive nuclear waste (HLW).

Reprocessing spent fuel is a technical process that recovers plutonium and reusable uranium from spent fuel and separates radioactive wastes into more technically manageable forms. The recovered plutonium is utilized into the nuclear power plants in the form of uranium-plutonium MOX fuel.

In addition, the highlight of Japanese nuclear policy so far has been deploying fast breeder reactors in order to efficiently consume uranium fuel. A FBR is a nuclear reactor which will produce more fissile nuclear material than it consumes. The breeding reaction in the FBR is to produce fissionable plutonium-239 (Pu239) from non-fissionable uranium-238. Uranium-238 is more abundant than fissionable uranium-235, which is consumed in typical light water reactors (LWR). The commercialization of FBRs in Japan was envisaged by 2050. The Joyo experimental FBR in Tokai, Japan has been operating successfully since 1977 and has accumulated a lot of operational experience and experimental data. The 280 MW Monju FBR started operation in 1994, but it was shut down due to a sodium leakage problem in its cooling system during experimental operation in 1995. In 2010, Monju was restarted as an experimental operation, but it shut down again due to a mechanical accident within the reactor.

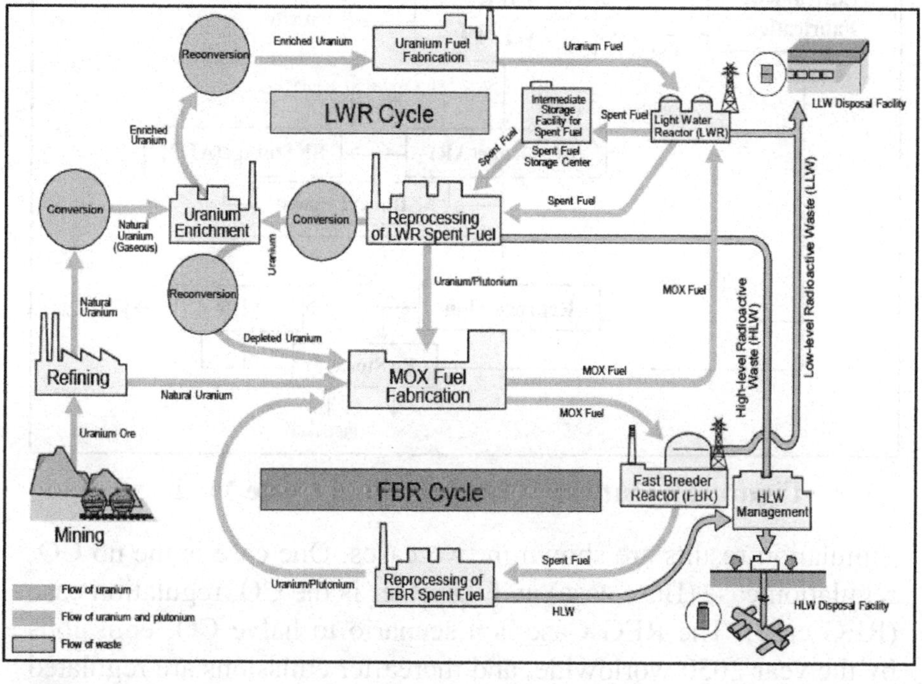

Figure 22. The Concept of Closed Nuclear Fuel Cycle in Japan.[40]

The author analyzed the impact of a closed nuclear fuel cycle (Figure 23) on energy markets in Asia and the world to 2100.[41] This

40. Federation of Electric Power Companies of Japan.

41. Saurabh Sharma, Ryoichi Komiyama and Yasumasa Fujii, "Assessment of Sustainable Energy Strategy with Long Term Global Energy Model Incorporat-

fuel cycle is expected to drastically improve the usage efficiency of natural uranium resources. Figure 23 shows the outlines of the nuclear fuel cycle model. Light-water reactors, light-water MOX reactors (LWR-MOX), and fast breeder reactors are considered as specific kinds of nuclear power generation technologies.

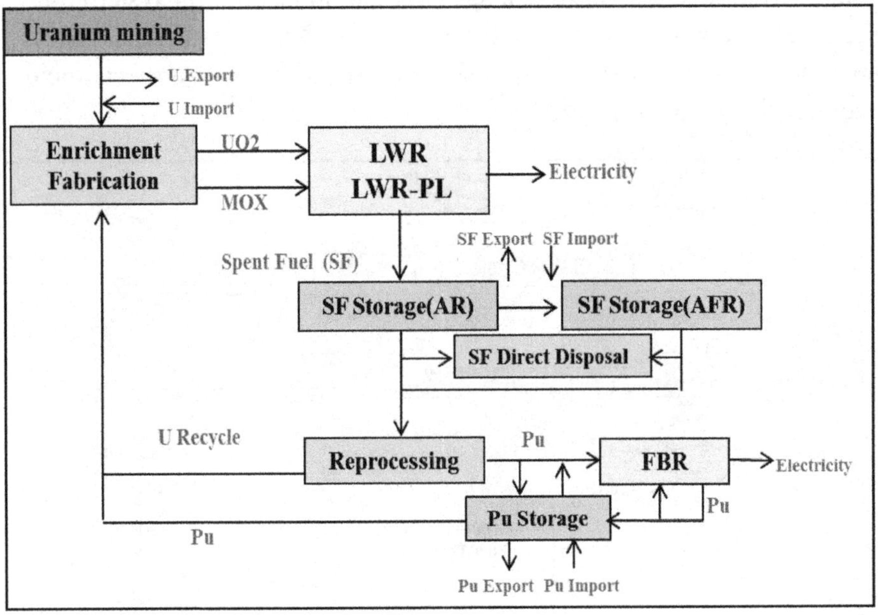

Figure 23. Outline of Nuclear Fuel Cycle Model.[42]

Simulation results are shown in two cases. One case is the no CO_2 regulation case (Base case) and the other is the CO_2 regulation case (REG case). The REG Case is a scenario to halve CO_2 emissions by the year 2050 worldwide, and thereafter emissions are regulated so that atmospheric CO_2 concentrations are maintained at the level needed to avoid around a 2° Celsius increase of average global temperature from pre-industrial levels. Furthermore, in the REG case, the developed countries (high-income countries) are assumed to reduce CO_2 emissions by 80% compared with levels in 2000.

ing Nuclear Fuel Cycle," *Journal of Environmental Science and Engineering B* 1, no. 11, November 2012, pp. 1215-1232.

42. Ibid.

Figures 24 and 25 show electric power generation and primary energy supply, respectively.[43] In the Base case, a majority of the world's primary energy is almost exclusively provided by coal, gas, and oil until the middle of the century. Particularly, coal, of which reserves and resources are abundant and economically affordable, shows remarkable growth of supply among fossil fuels. After the middle of this century, with the peak in extraction of conventional sources, unconventional oil and gas, which are more expensive than conventional ones, will start to be produced. This economic deterioration of fossil fuels partly encourages the introduction of nuclear energy and, to a lesser extent, renewable energy such as solar, biomass, and wind power. This fossil fuel-intensive scenario leads to substantial CO_2 emissions growth.

By contrast, in the REG case, the imposition of a carbon regulation target encourages the large scale adoption of low carbon energy. Around the world, at the beginning of the century, coal, concentrated in thermal plants, becomes significantly less competitive due to carbon penalties, although integrated gasification combined cycle (IGCC) with carbon capture and storage (CCS) plays an important role later in the century. On the other hand, natural gas is introduced early in the century based on its economic attractiveness, and maintains this position later with the adoption of CCS, with gas-fired power plants supplying around a quarter of total electric power capacity in the second half of the century. Concerning nuclear energy, nuclear LWR usage is limited in the second half of the century by exhaustion of uranium resources. Introduction of FBRs enables these technologies to supply power requirements well beyond 2050. In addition, achieving low stabilization does not appear to be possible without large-scale deployment of renewables over the long term. Later in the century, biomass, solar, and wind power are expected to play an essential role in decarbonizing electric power supply. The results of the REG case reveal that there is no single technology to realize a sustainable energy pathway in the long-term perspective.

43. Ibid.

The result indicates that nuclear power plants with nuclear fuel recycling are estimated to play significant roles to reduce CO_2 emissions. With a great deal of uncertainty, it is difficult to draw firm conclusions on which options have the largest potential in achieving significant CO_2 reductions. However, the simulation results indicate that massive CO_2 mitigation targets can be satisfied with the large-scale deployment of innovative technology, highlighting essential roles for nuclear, renewables, efficient use of fossil fuels, and CCS. The results provide support for simultaneously pursuing multiple innovative technologies, instead of concentrating on a single technology and focusing on realistic technological measures on the basis of current technological understanding.

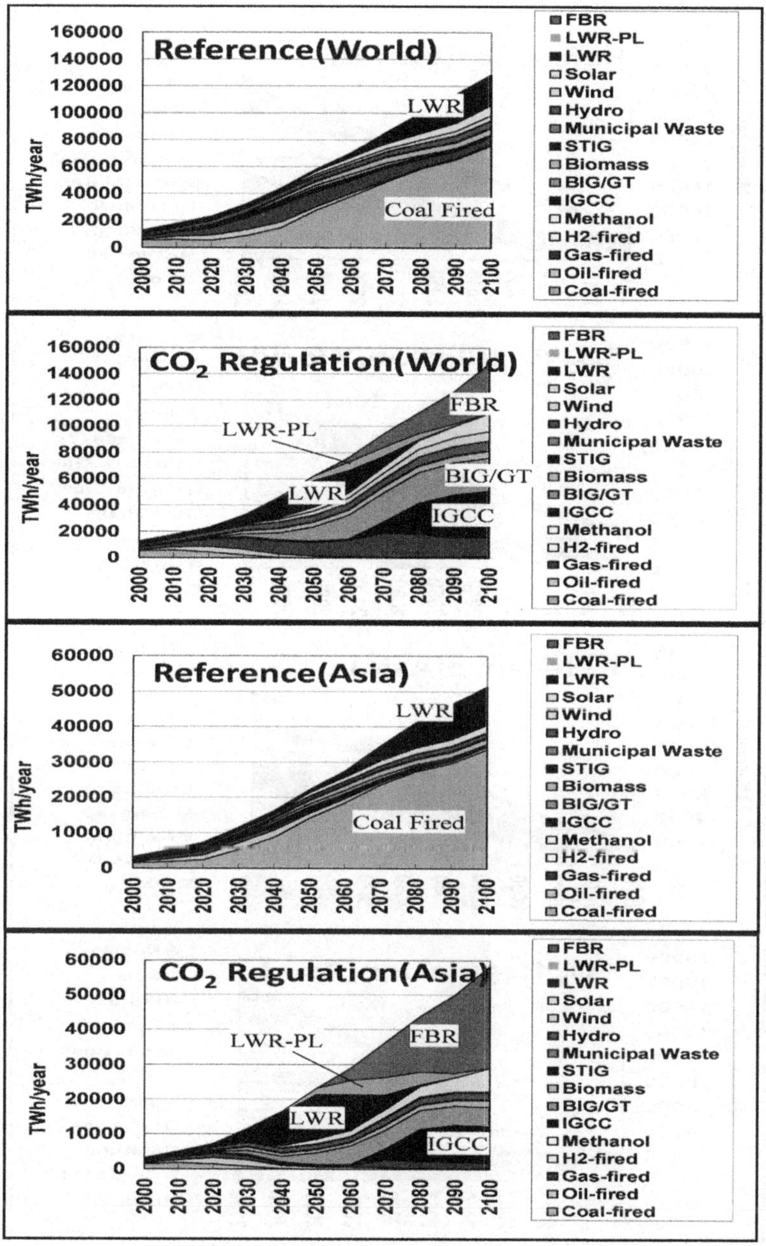

Figure 24. Power Generation Mix to 2100 in the World and Asia.[44]

44. Ibid.

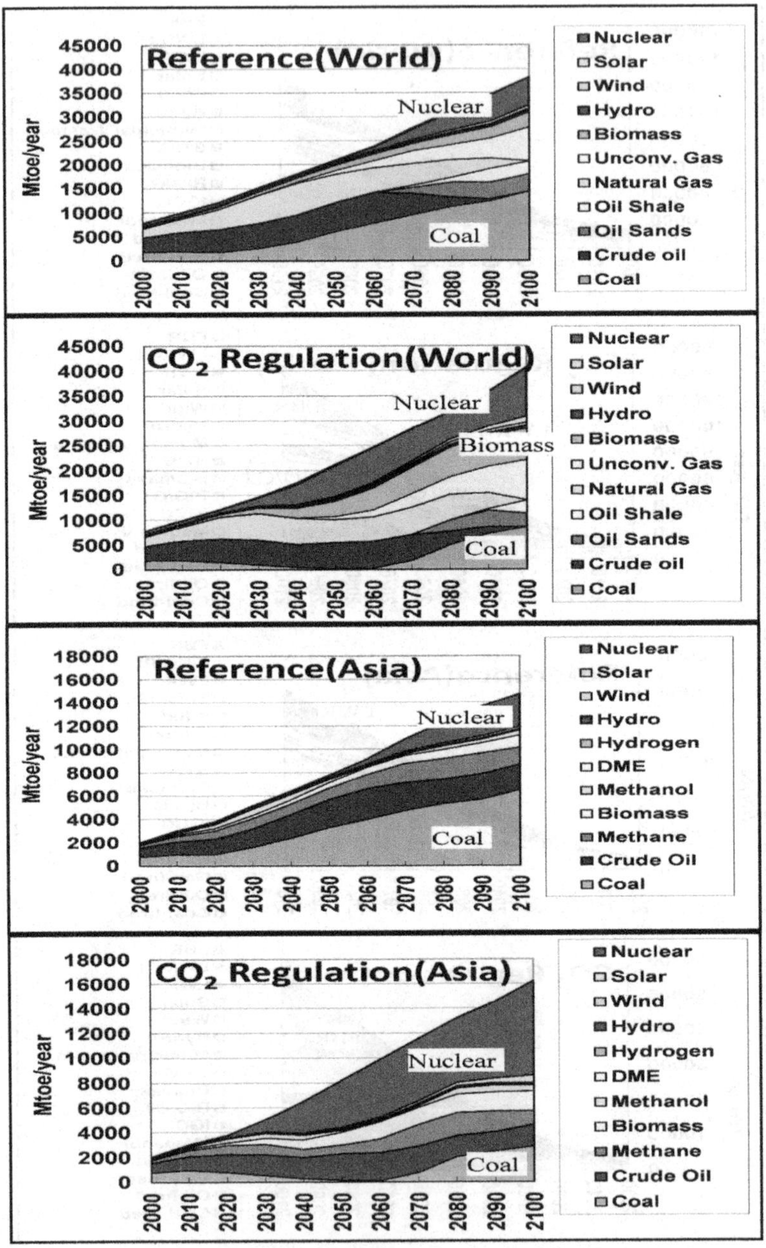

Figure 25. Primary Energy Supply to 2100 in the World and Asia.[45]

45. Ibid.

Concerning the nuclear perspective, continuous use of nuclear energy is impossible if LWR is the only technology utilized and the closed nuclear fuel cycle is unavailable through this century, because of the depletion in uranium resources. However, FBRs, which consume the plutonium included in spent fuel, enables these technologies to supply the power requirements well beyond 2050. Thus, nuclear proves to be potentially critical for creating sustainable pathways in the medium to long term future if the closed nuclear fuel cycle and FBRs become commercially viable options. Large-scale application of renewables and CCS are observed to be key technologies in the mitigation scenario, particularly in the long term.

Conclusions

Balancing Asia's rapid energy demand growth in a sustainable way is a problem to be urgently resolved. However, there is no single panacea to solve the energy and environmental issues and it is more desirable to pursue multiple innovative technologies instead of focusing only on realistic technical measures on the basis of current technological understanding. In this regard, nuclear power plants are an indispensable technology even after Fukushima to tackle energy and environmental concerns. Nuclear energy accounts for a small portion of the Asian energy supply and its deployment will help to diversify the Asian energy mix. Such diversification is necessary for energy security reasons because the current heavy reliance on fossil energy such as coal in the Asian region is not sustainable for the world. Additionally, reducing greenhouse gas emissions requires the deployment of clean alternative energy to replace thermal power generators. Provided that renewable energies such as PV and wind are unable to generate power at a higher intensity and on a large-scale basis, nuclear energy is the only technically feasible option for replacing fossil fuels.

However, current LWR technology is unable to provide a sustainable supply of nuclear energy in the long-term. Despite this, nuclear is expected to play a significant role in resolving energy security and environmental issues. In order to mitigate the exhaustion of uranium resources and to continue the deployment of nuclear technology, the closed nuclear fuel cycle with fast breeder reactors are indispensable options in terms of saving uranium resources and reducing the amount of high-level radioactive waste. Utilization of only LWRs as a nuclear power source accelerates the depletion of uranium resources and rapidly increases the high-level radioactive waste, which is currently one of the major impediments to nuclear energy. Therefore, continuous R&D efforts on advanced nuclear technology are necessary to accomplish the sustainable pathway, particularly in Asia where rapid electricity demand is projected and the massive deployment of nuclear technology is expected. Japan has the most technical and operational experiences with nuclear power plants and fast breeder reactors in Asia. Based on its extensive experience with nuclear power plants and the stringent nuclear safety regulations intensified after Fukushima, Japan should aim to develop more advanced nuclear technology and policy with higher safety, security, and safeguards as quickly as possible, which should be shared among Asian countries. Japan should effectively utilize these advanced nuclear technologies and policies to support Asian countries in achieving sustainable growth.

CHAPTER 6

Nuclear Development in East Asia: A Chinese Perspective

Gu Zhongmao

Challenges to East Asian Energy Security

The three major East Asian countries, China, Japan, and South Korea, have made great contributions to the world economy and at the same time consume large amounts of energy. The primary energy consumption of these three countries, according to a 2012 BP report, accounts for some 28% of the world total.[1] However, these East Asian countries have their own inherent energy security problems. Both Japan and South Korea lack fossil energy resources. Japan and South Korea's primary energy dependence on overseas sources is more than 96% and 97% respectively.[2]

China's energy security is also facing two major challenges. The first one is the contradiction between the country's ever increasing demands for energy and its insufficient reserves of fossil fuels. As a developing country, China's per capita energy consumption has just

---

1. *BP Statistical Review of World Energy June 2012*, London: BP, June 2012, p. 40, available from *www.bp.com/content/dam/bp/pdf/Statistical-Review-2012/statistical_review_of_world_energy_2012.pdf.*

2. Yusuke Kuno, "Nuclear Future and Nuclear Non-Proliferation in Japan," presented at the workshop on "A Regional Network of Nuclear Fuel Cycle Facilities for the Asia Pacific," Tokyo, Japan, March 2009, pp. 6-17 and H.G. Lee, "Korea's nuclear energy policy and programs: past, present and future," presented at "The China-Korea Workshop on Energy Science," Beijing, China,May 6-8, 2010.

reached the world average. China's energy supply will continue to grow significantly in the coming decades to support the further development of its national economy. The total amount of coal reserves in mainland China is large, ranking third in the world. However, China's per capita coal reserve is much lower than the world average. China is also short of oil and gas reserves. With the drastic increase of oil and gas consumption since the mid-1990s, China is becoming more and more dependent on imports of oil and gas from abroad.

The second challenge comes from the irrational energy structure, in which coal provides 60% of primary energy and 77% of power generation. The coal-based energy structure causes serious problems of environmental pollution and carbon emissions. Acid rain covers more than one-third of the country's land and haze has become significantly heavier in recent years. The economic loss caused by environmental pollution accounts for about 3-7% of gross domestic product (GDP) in China. Since 2010, China's GHG emissions have exceeded the United States and now ranks first in the world, though China's per capita discharge of GHGs is only less than one-fourth of that of the United States. These problems will hinder the further development of the country if they are not solved properly.

Meanwhile, China's energy efficiency is very low. Heavy industry with high energy consumption and high pollution dominates, consuming some 50% of the total energy supply. The primary energy consumption per unit of GDP is 8.7 times and 2.7 times of Japan and South Korea respectively.

It is therefore commonly accepted that China needs to place the development of energy saving technologies as its first priority. And, it has to adjust from the current energy and industrial structure in place, gradually reducing the share of high pollution energy resources and high energy-consuming industries.

To cope with the serious problems of environmental pollution and global warming, alternative energy, such as wind and solar energies, have received more and more attention. In China, hydro power resources are abundant. China's economically exploitable hydro power resources are estimated to be some 400 gigawatts electrical (GWe), of which the installed capacity of hydro power has reached some 280 GWe by 2013.[3] China will continue to develop hydro power as a top priority in the coming decades.

Non-hydro renewable resources, such as wind, solar, and biomass energies, are highly encouraged. For example, the installed capacities of wind and solar power in 2013 reached 75.5 GWe and 14.8 GWe, respectively, and are expected to be increased to 200 GWe and 160 GWe respectively by 2020.[4]

However, it seems that renewable energy, such as wind and solar, will not be able to replace fossil fuels as a base load source in the foreseeable future owing to their very low energy density, low availability factors, and relatively high costs.

Nuclear Development Plans in East Asian Countries and the Impact of Fukushima

To ensure energy security, both Japan and South Korea have paid great attention to the development of nuclear power since the 1970s. The nuclear share in these two countries had reached 25% and 36% respectively before the Fukushima nuclear accident.

As a well-developed clean energy with near-zero carbon emissions, nuclear energy has received more and more attention in China since the 1990s and is regarded as a supplement to the fossil fuels together with other new energies. There is much room for the development of nuclear energy in China.

---

3. Wang Jing, "Electricity policy and impact in China," presented at the workshop on "The Future of Nuclear Power in China and the World: The Policy Context," *Carnegie-Tsinghua Global Policy Center*, Beijing, China, September 11-12, 2014.

4. "China's energy structure adjustment enters spurt period," *China Environmental Protection*, September 28, 2014.

Nuclear Development Plans in East Asian Countries before the Fukushima Nuclear Accident

In Japan, 54 commercial light water reactors (LWRs) (48.9 GWe) were in operation, supplying about 25% of total electricity generation before the Fukushima nuclear accident in 2011.[5] According to the original plans of the electric utilities, a total of 66 LWRs (65.1 GWe) would be operating in 2020.[6] South Korea has 23 units, including 19 light water pressurized water reactors (PWRs) and four heavy water reactors (HWRs) that were in operation in 2012.[7]

The Korean National Energy Committee, the top policymaking body for energy in South Korea, adopted a new National Energy Basic Plan (NEBP) on August 28, 2008, which proposed that the percentage of electricity generated by nuclear power should increase to 59% by 2030 from the current 36%.[8] This plan was approved by the South Korean government at the 255th Atomic Energy Commission (AEC) meeting held on December 22, 2008, during which the "Long-term Development Plan for Future Nuclear Energy System" was announced. As of 2010, South Korea had 20 nuclear units (17.7 GWe) in operation and eight units (9.6 GWe) under construction. According to the national plan, a total of 32 reactors will be operat-

5. Tomoko Murakami, "Japan's Nuclear Power Development Strategy and Possible Regional Cooperation—Post Fukushima," presented at the workshop on "Global New Energy Development Strategies and Northeast Asia Regional Cooperation," Beijing, China, June 24, 2011.

6. Tatsujiro Suzuki, "Status and Plans of Nuclear Energy for Fuel Cycle in Japan," presented at 5th Annual Workshop on "Nuclear Energy Non-Proliferation in East Asia," Busan, Republic of Korea, June 28–July 1, 2009.

7. Hansoo Lee, "R&D Activities of Nuclear Fuel Cycle in Korea," presented at the 10th Korea-China Joint Workshop on "Nuclear Waste Management and Fast Reactor Technologies," Busan, Republic of Korea, October 10-11, 2012.

8. Won Il Ko and Hansoo Lee, "Status of Spent Fuel Management and System Analysis of Nuclear Fuel Cycle in Korea," presented at the 10th Korea-China Joint Workshop on "Nuclear Waste Management and Fast Reactor Technologies," Busan, Republic of Korea, October 10-11, 2012.

ing by 2022 and a total of 40 reactors will be in operation by 2030.[9]

In China, 13 nuclear units (11 GWe) were in operation with 34 units (36.9 GWe) approved as of September 2010. China announced that by 2020 nuclear power capacity will be increased to 70 GWe in operation with another 18 GWe under construction.[10]

Impact of the Fukushima Nuclear Accident on Global and East Asian Nuclear Energy

The Fukushima nuclear disaster on March 11, 2011, was a great shock to the global nuclear industry. Sentiments for phasing out nuclear energy ran high immediately after Fukushima. Germany announced its intention to close all its nuclear power plants (NPPs) by 2022. Facing strong public opposition, Japan had to close most of its NPPs.

However, the Fukushima nuclear accident has not changed the challenging global population, resources, and environmental problems. There is clear evidence of changes in the composition of the GHGs in the lower atmosphere, with carbon dioxide (CO_2) in particular steadily increasing to its present level of about 390 parts per million (ppm). It has increased by one-third in the last 200 years, and half of that in the last 30 years as shown in Figure 1.

9. Dong-Wook Jerng, "Perspectives on the Nuclear Energy Program and R&D Directions in Korea," presented at the 6th International Workshop on "Nuclear Energy and Non-Proliferation in East and Southeast Asia," Hyundai Hotel, Gyeongju, Republic of Korea, October 27-29, 2010.

10. Gu Zhongmao, "Envision of China's Nuclear Energy and Fuel Cycle Development—Geared to International Trend," presented at the workshop on "Responsibility for Nuclear Suppliers," Beijing Capital Hotel,China, July 18, 2013.

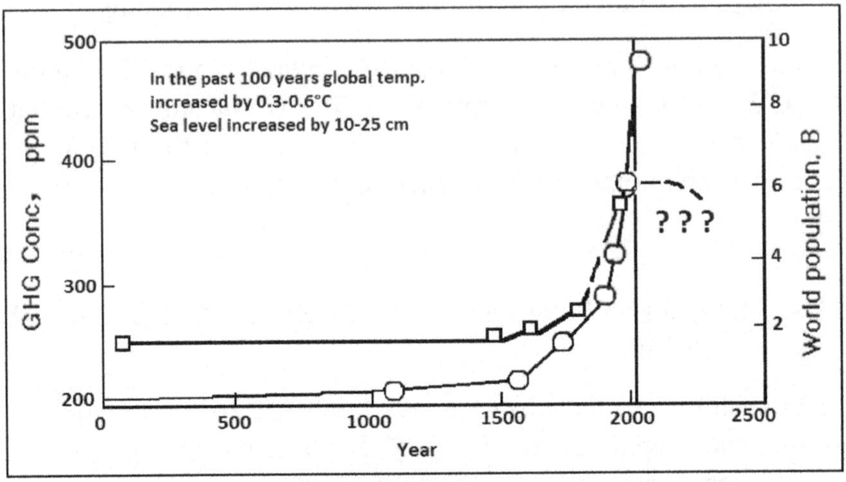

Figure 1. World population and GHGs Concentration vs Time.[11]

The global renaissance of nuclear energy that occurred at the beginning of this century can be mainly attributed to coping with global warming. Worldwide nuclear power, with the installed capacity of some 370 GWe, contributes to CO_2 reduction of 2.2 billion metric tons annually (t/a), constituting an 8% reduction of the global greenhouse gas emissions. It is the author's understanding that nuclear power should not be "phased out" unless our carbon reduction commitment could be "phased out." In fact, the shut-down of NPPs in Japan caused a 5.8% increase in CO_2 emissions since 2011.[12]

11. This figure was plotted with the data cited from the following references: World Energy Council, *Energy for Tomorrow's World – Acting Now*, London: World Energy Council, 2000; "ANS Winter Meeting: Nuclear and energy independence," *Nuclear News*, January 2002, pp. 40-43, 46-58, available from *www2.ans.org/pubs/magazines/nn/docs/2002-1-3.pdf* and B.L. Eyre "Power Generation for the Twenty-First Century: What is the Role for Nuclear?" *Nuclear Energy* 37, no. 1, 1998, pp. 59-72.

12. Tatsujiro Suzuki, "Nuclear Energy and Nuclear Fuel Cycle Policy Options After the Fukushima Accident," presented at the Nonproliferation Policy Education Center's "East Asian Alternative Nuclear Weapons Futures Conference" and "East Asian Alternative Energy Futures Conference," Honolulu, Hawaii, Febru-

In China, immediately after Fukushima nuclear accident, the State Council made four policy decisions on March 16, 2011: Immediate safety reviews of all nuclear facilities; strengthening the safety management of operational facilities; comprehensive reviews of the nuclear facilities under construction; and suspending the approval of the new projects until after new nuclear safety plans are issued.

The general safety examination showed that NPPs both in operation and under construction met the requirements of China's present safety criteria.

On October 24, 2012, the State Council made the decision to restart the nuclear power program with safety and high efficiency as priorities. The goal of nuclear power capacity by 2020 was adjusted to 58 GWe in operation and 30 GWe under construction. However, no new nuclear power project has been approved by the government since the State Council decision was made in 2012. One of the possible reasons may be the debate over which reactor type should be selected in deploying third generation NPPs in China. Recently, China has decided to build two third generation demonstration projects; two units of the Hualong 1 design (1000 MWe capacity) in Fujian and Guangxi provinces respectively, and two units of the CAP1400 design (1400 MWe) in Shandong province.[13]

In South Korea, it seems that the Fukushima nuclear accident has had no big impact on the nuclear energy program. South Korea continues to follow the long-term national plan of nuclear energy development set in December 2008.

Japan has suffered greatly from the Fukushima nuclear accident. The post-accident clean-up is still under way and will possibly be extended for decades. As of June 2011, more than 80% of Japa-

---

ary 26, 2014.

13. "Hualong One: China's Nuclear Power Embarks on a New Journey," *ChinaPower.com*, September 23, 2014.

nese say they are anti-nuclear.[14] As of March 27, 2012, Japan had only one out of 54 nuclear reactors operating: The Tomari-3, after the Kashiwazaki-Kariwa 6 was shut down.[15] The Tomari-3 was shut down for maintenance on May 5, 2012, leaving Japan with no nuclear-derived electricity for the first time since 1970.[16] Despite protests, on July 1, 2012, unit three of the Ōi Nuclear Power Plant was restarted. In September 2013, Ōi units three and four went offline, making Japan again completely without nuclear-produced electrical power. On August 11, 2015, the Sendai Nuclear Power Plant was brought back online, followed by two units (three and four) of the Takahama Nuclear Power Plant on January 29, 2016. However unit four was shut down three days after restart due to an internal failure and unit three in March 2016 after district court in Shiga prefecture issued an injunction to halt operation of Takahama Nuclear Power Plant.[17] However, some experts in Japan note that, as a resource-lacking country, Japan needs nuclear energy for their energy security even after the accident.[18]

In February 2013, Japanese Prime Minister Shinzo Abe announced that Japan will restart NPPs where safety has been confirmed and

---

14. M. V.Ramana, "Nuclear power and the public,"*Bulletin of the Atomic Scientists* 67, no. 4, July 2011, p. 43.

15. Monica Kleja, "Soon Japan's nuclear power stands still," *NyTeknik* (New Technology), March 27, 2012, translation available from *translate.google.dk/translate?hl=da&sl=sv&tl=en&u=http%3A%2F%2Fwww.nyteknik.se%2Fnyheter%2Fenergi_miljo%2Fkarnkraft%2Farticle3436603.ece&sandbox=1.*

16. David Batty, "Japan shuts down last working nuclear reactor",*The Guardian*, May 5, 2012, available from *www.theguardian.com/world/2012/may/05/japan-shuts-down-last-nuclear-reactor*.

17. "Nuclear power in Japan," Wikipedia, accessed on October 26, 2016, available from *https://en.wikipedia.org/wiki/Nuclear_power_in_Japan.*

18. Endo Tetsuya, "The International Impact of the Fukushima Nuclear Plant Accident and Lessons To Be Learned," presented at the Peking University workshop on "Global New Energy Development Strategies and Northeast Asia Regional Cooperation," Beijing, China, June 24, 2011.

promote the introduction of energy conservation and renewable energies to the greatest possible extent to reduce the degree of dependency on nuclear power as much as possible.[19]

In February 2014, the draft of the Japanese Basic Energy Plan suggested that nuclear energy is still among the country's vital sources of electricity. Under the plan, Japan's nuclear energy dependency will be reduced as much as possible, but reactors meeting the new safety standards set after the 2011 nuclear crisis should be restarted. The draft also said that a mix of nuclear, renewable, and fossil fuels will be the most reliable and stable source of electricity to meet Japan's energy needs.[20] It was reported that on April 11, 2014, Japan's cabinet approved the Basic Energy Plan, in which nuclear energy was defined as an important base-load energy source.[21]

Issues of Sustainability of Nuclear Energy in East Asia

Spent Fuel Accumulation in East Asia

In Japan, 23,600 metric tons (t) of spent fuel had been generated by 2008. It is expected that another 40,000 t of spent fuel will be generated from 2009 to 2049 according to the original plan set before 2011.[22] There will be some uncertainty of Japan's nuclear power capacity in the coming decades.

---

19. Suzuki, "Nuclear Energy and Nuclear Fuel Cycle Policy Options After the Fukushima Accident."

20. "Nuclear is key energy source, Japan's first energy policy draft since Fukushima says," *RT News*, February 25, 2014, available from *rt.com/news/japan-nuclear-energy-source-675/*.

21. Martin Fritz, "Japan reverses its withdrawal from nuclear power," *Deutsche Welle*, April 13, 2014, available from *www.dw.de/japan-reverses-its-withdrawal-from-nuclear-power/a-17563405*.

22. Suzuki, "Status and Plans of Nuclear Energy for Fuel Cycle in Japan."

In South Korea, 14,000 t of spent fuel were generated by 2008. It is expected that the total amount of spent fuel generated will be 50,000 t by 2050.[23]

In China, about 2,500 t (Mainland) plus 3,000 t (Taiwan) of spent fuel have been generated from NPPs presently. It is estimated that by 2020, with ca. 60-80 GWe of nuclear power in operation, spent fuel accumulation would be ca. 10,000 t. By 2030, assuming nuclear capacity reached 130 GWe, spent fuel could reach ca. 30,000-40,000 t. It may be possible that the PWRs' capacity could reach up to 200 GWe by 2050 and there could be possibly ca. 90,000 t of spent fuel generated.

It is envisioned that the total amount of spent fuel generated in East Asia may reach some 200,000 t by 2050, which will contain some 2,000 t of plutonium (Pu). The build-up of this huge amount of spent fuel will be a large burden for waste storage and disposal, and could possibly cause increased proliferation risks in East Asia.

Sustainability Considerations: Arguments on Two Options of the Nuclear Fuel Cycle

Over the past decades, the debate on two nuclear fuel cycle options, the once-through cycle (OTC) and the closed fuel cycle (CFC), have been continuous among the global community and no consensus has been reached. Countries such as France, the United Kingdom, Russia, India, Japan, South Korea, and China have pursued the CFC option while Sweden, Switzerland, Finland, Canada and the United States selected the OTC option. The sustainability of nuclear energy is the major issue to be considered when choosing a fuel cycle option.

---

23. Y.S. Hwang and K.C. Song, "Status and Plans for Nuclear Energy and Fuel Cycle in the Republic of Korea," presented at 5th Annual Workshop on "Nuclear Energy Non-Proliferation in East Asia," Busan, Republic of Korea, June 28-July 1, 2009.

Sustainable nuclear energy must contribute to the well-being of future generations by reducing the use of natural resources and avoiding detrimental effects on public health and the environment. In addition to ensuring safety, economic competitiveness and non-proliferation (i.e. protection against diversion or undeclared nuclear material production and misuse of technology), a sustainable nuclear fuel cycle must also be linked to the following two issues: Maximum utilization of uranium resources and minimization of nuclear waste and its impact on the environment.

The OTC option involves direct disposal of the spent fuel after interim storage for cooling when it is then packaged (see Figure 2). The OTC option is simpler and more economic at present uranium prices, compared with the CFC option. Meanwhile, it is thought that the high radiation emissions of spent fuel can prevent proliferators from gaining access and thus the OTC option is favored for nonproliferation.[24]

Concerning the secured supply of uranium resources, it is reported by the Organization for Economic Co-operation and Development's *Red Book 2012* that global conventional uranium resources are 15.40 megatons or million metric tons (Mt), which includes some 5.30 Mt of identified conventional resources and some 10 Mt of undiscovered conventional resources. These conventional uranium resources can support the present global nuclear capacity (373 GWe) in the OTC case only for less than 100 years as shown in Figure 3. Studies showed that no significant differences would result in the case of MOX fuels use in PWRs.[25]

24. Frank Von Hippel and Lora Saalman,"Plutonium, Nonproliferation, and Radioactive Waste Politics in East Asia: Why the once-through nuclear fuel cycle is the most economical and proliferation-resistant," presentation at Peking University, Beijing, China, November 1, 2010.

25. Vincenzo Romanello, Massimo Salvatores, et al., "Sustainable Nuclear Fuel Cycles and World Regional Issues," *Sustainability* 4, no. 6, June 2012, pp. 1214-1238.

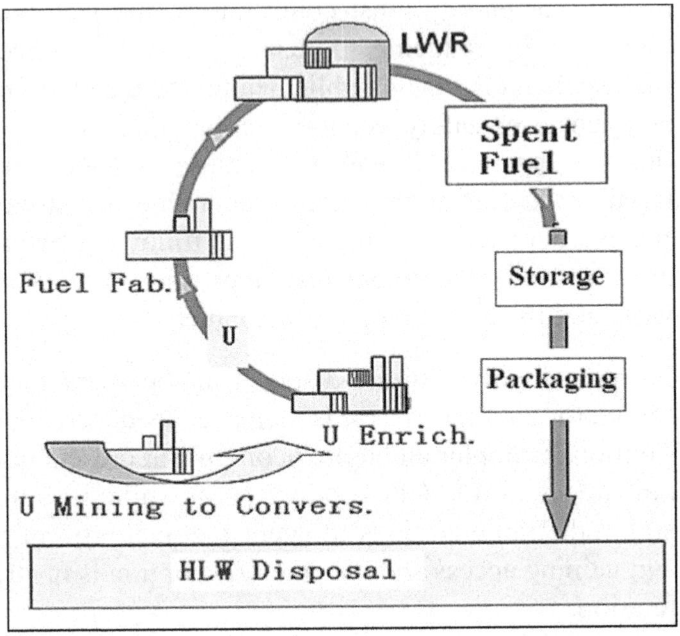

Figure 2. Schematic of Once-through Cycle (OTC).

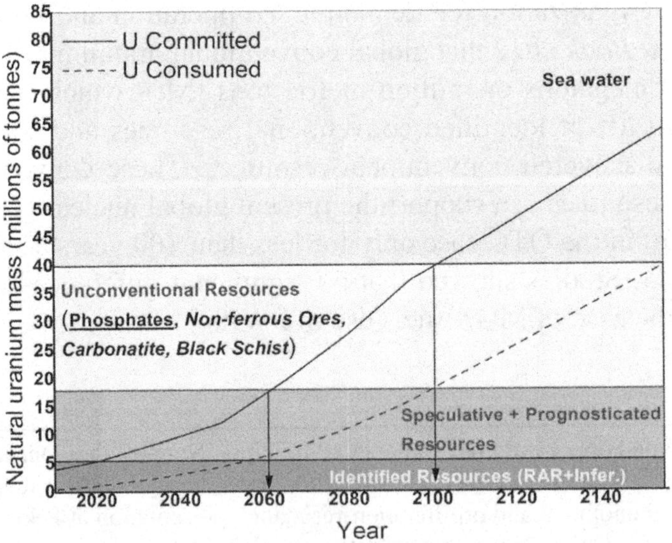

Figure 3: Natural Uranium Consumed and Engaged for PWRs OTC Case.[26]

26. Ibid.

As a consequence of utilizing an OTC option for the world fuel cycle, a large amount of spent fuel will accumulate worldwide. It is expected that more than 800,000 t of spent fuel will be generated globally by 2150, which will contain ca. 8,000 t of plutonium.[27] Such a large amount of worldwide spent fuel will cause serious problems of geologic disposal. Where are we to find and how will we construct the geologic repositories large enough for disposing of these spent fuels? This will be a big burden to be transferred to future generations. It is obvious that the OTC case is not a sustainable option.

As for the issue of nonproliferation, after storing the spent fuel for longer periods of time (e.g. 100-200 years), the radiation level of spent fuel will become weaker and not be adequate to stop proliferators from accessing the spent fuel. Finally, the spent fuel could become a "plutonium mine." The uncertainty of the integration of spent fuel (especially claddings) after prolonged storage is another concern. Therefore, the prolonged storage of spent fuel is not a final solution. Rather, by utilizing reprocessing or recycling measures, plutonium can be burned in reactors. It is argued that the CFC option can eventually eliminate plutonium and will be essential for nonproliferation if International Atomic Energy Agency (IAEA) safeguards are fully implemented throughout the whole fuel cycle process.

In order to address potential future uranium resource shortages, a group of European experts analyzed different scenarios and envisioned a transition from a PWR-based OTC fuel cycle to an advanced CFC with fast reactors (FRs) for the future of nuclear energy.[28] They projected the nuclear energy production share in Asia, using PWRs and fast breeder reactors (FBRs), as shown in Figure 4.

27. Michael Weber, "Where will we be in 2050?" presentation at the U.S. Nuclear Regulatory Commission's "2010 Spent Fuel Storage and Transportation Licensing Process Conference," Rockville, MD, June 23, 2010.

28. Romanello, Salvatores, et al.

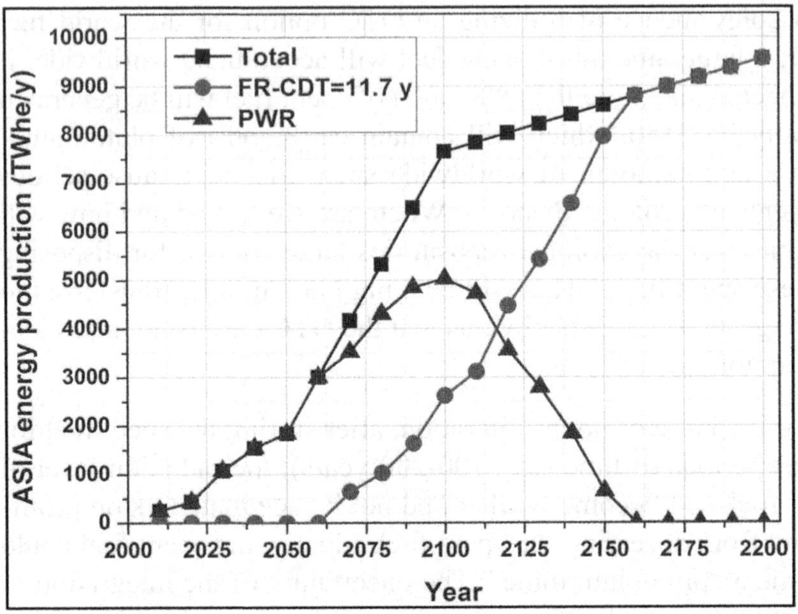

Figure 4. Projected Nuclear Energy Share in Asia (PWRs and FBRs).[29]

From the above analysis, one can see the reasons why most major nuclear energy countries selected a policy of developing advanced CFC with FRs and expect the gradual transition from PWRs to FRs.

It is worthwhile to note the evolution of the nuclear fuel cycle policy in the United States in the past decades. It is my understanding that the core of the U.S. nuclear energy policy is to maintain its world leader position and that nonproliferation is a key point of consideration.

During the 1950s until the early 1970s, U.S. nuclear policy was in favor of CFC. In the early 1950s, fearing the spread of nuclear weapons in the world, the United States encouraged research and development (R&D) on civil utilization of nuclear energy, which was marked by the famous speech "Atoms for Peace" delivered by President Dwight D. Eisenhower to the United Nations General Assembly on December 8, 1953. The United States then launched an

29. Ibid.

"Atoms for Peace" program that supplied nuclear energy and fuel cycle technology and information throughout the world, and the United States pioneered reprocessing activities as well. As a result, "Atoms for Peace" opened up nuclear research to civilians and countries that had not previously possessed nuclear technology. This made it possible for some countries to develop weapons under the cover of peaceful nuclear energy use. Under "Atoms for Peace" related programs, the United States exported over 25 tons of highly enriched uranium (HEU) to 30 countries, mostly to fuel research reactors, which are now regarded as proliferation and terrorism risks.[30]

On May 18, 1974, India announced its first nuclear test, described as a "peaceful nuclear explosion." India's nuclear test used plutonium separated with U.S.-provided technology. This action caused the United States to reverse its policy from favoring CFC to favoring OTC. In 1976, President Gerald Ford announced that the reprocessing and recycling of plutonium should not proceed. In 1977, President Jimmy Carter announced that the United States would defer the commercial reprocessing and recycling of plutonium indefinitely.

By the end of the 1990s, more than 500,000 t of spent fuel had accumulated in the United States as a result of its OTC policy, which hindered the further development of nuclear energy in the United States, which was losing its world leader position in nuclear energy. The U.S. Department of Energy (DOE) then supported the Accelerator Transmutation of Waste (ATW) program to separate plutonium and minor actinides from spent fuel and burn them in fast burner reactors. The ATW program was re-named as the Advanced Fuel Cycle Initiative (AFCI) program in 2002. These subtle changes can be regarded as the discontinuity of the U.S. OTC policy.[31]

---

30. Jessica C. Varnum, "60 Years of Atoms for Peace," *Nuclear Engineering International*, January 23, 2014, available from *www.neimagazine.com/features/feature60-years-of-atoms-for-peace-4164653/*.

31. H. AitAbderrahim, et al., "Accelerator and Target Technology for Accelera-

In 2006, the Administration of President George W. Bush reversed the OTC policy by putting forward the Global Nuclear Energy Partnership (GNEP) Initiative. The GNEP Initiative focused on trying to develop proliferation-resistant reprocessing and fast burner reactors in order to alleviate the spent fuel burden, to restrict the fuel cycle technologies in limited countries for nonproliferation purposes, and to restore the world leader position of the United States in nuclear energy.[32]

In 2009, however, the Administration of President Barack Obama negated the Bush Administration's nuclear energy policy and restored the OTC policy again. From the dramatic development (CFC-OTC-CFC-OTC) of U.S. nuclear fuel cycle policy in the past 50 years, one can expect further changes to the U.S. fuel cycle policy in the coming decades.

Proposed Programs of Reprocessing/Recycling in East Asia

Japan

Japan is the only non-nuclear weapon state in the world that has been allowed by the United States to pursue a reprocessing program.[33]

Since the 1970s, Japan has pursued an ambitious fast breeder program. In 2009, Japan re-organized its R&D programs with newly defined targets for its demonstration by around 2025 and commer-

---

tor Driven Transmutation and Energy Production," September 17, 2010, available from *science.energy.gov/~/media/hep/pdf/files/pdfs/ADS_White_Paper_final.pdf.*

32. John Herczeg, "Global Nuclear Energy Partnership: Demonstration Projects," presented at a meeting of the China Commission of Development and Reformation, Beijing, China, May 22, 2006.

33. Charles S. Costello III, "Nuclear Non-proliferation: A Hidden but Contentious Issue in US-Japan Relations During the Carter Administration (1977-1981)," *Asia-Pacific Perspectives* III, no. 1, May 2003, pp. 1-6.

cialization by 2050, as set by the government's "Nuclear Power Nation Plan."[34]

Teksou Fukasawa of Hitachi-General Electric Nuclear Energy, described the present status of the nuclear fuel cycle system in Japan as shown in Figure 5. According to Fukasawa, the Rokkasho Reprocessing Plant (800 t/a) has completed active tests. The mixed-oxide (MOX) fuel fabrication plant (130 t/a) will start a hot test in 2015-2016. Because the spent fuel storage pool at Rokkasho (3000 t storage capacity) is almost full, discussions about building a second reprocessing plant, which is necessary for FBR deployment, should start in the coming years. Owing to the Fukushima nuclear accident, the FBR cycle technology (FaCT) development is delayed. The transition from LWR to FBR will be possibly starting between 2050 and 2100 and last for at least 60 years.[35]

34. Suzuki, "Status and Plans of Nuclear Energy for Fuel Cycle in Japan."

35. Tetsuo Fukasawa, "Flexible Fuel Cycle Initiative to Cope with the Uncertainties after Fukushima Daiichi NPP accident," presented at the 2nd China-Japan "Academic Symposium on Nuclear Fuel Cycle," Shanghai, China, November 27-30, 2013.

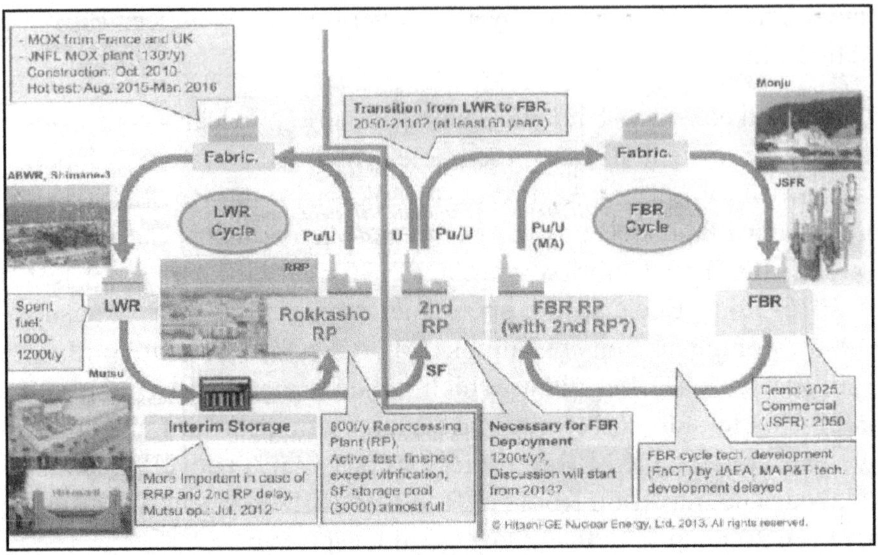

Figure 5. Nuclear Fuel Cycle Status in Japan.[36]

Some Japanese experts believe that nuclear energy is necessary for Japan even after the Fukushima nuclear accident, but that the transition from PWRs to FBRs will be delayed. Facing the post-Fukushima uncertainties in fuel cycle, a flexible fuel cycle initiative (FFCI) system is proposed as shown in Figure 6.[37]

The FFCI system removes most uranium (U) from the LWR spent fuel and stores the residue recycle material (RM), which contains Pu, residual uranium and minor actinides (Mas), and fission products (FPs), for future FBRs utilization in case of FBR cycle delay. If the FBR cycle develops smoothly, the separated RM could enter the FBR cycle directly. Using the innovative technology, 90% uranium could be removed from LWR spent fuel. The volume of the residual RM is expected to be reduced to ca. 1/15 of that of the spent fuel and the higher radiation level of RM compared to spent fuel could make it more proliferation-resistant.[38]

36. Ibid.

37. Ibid.

38. Ibid.

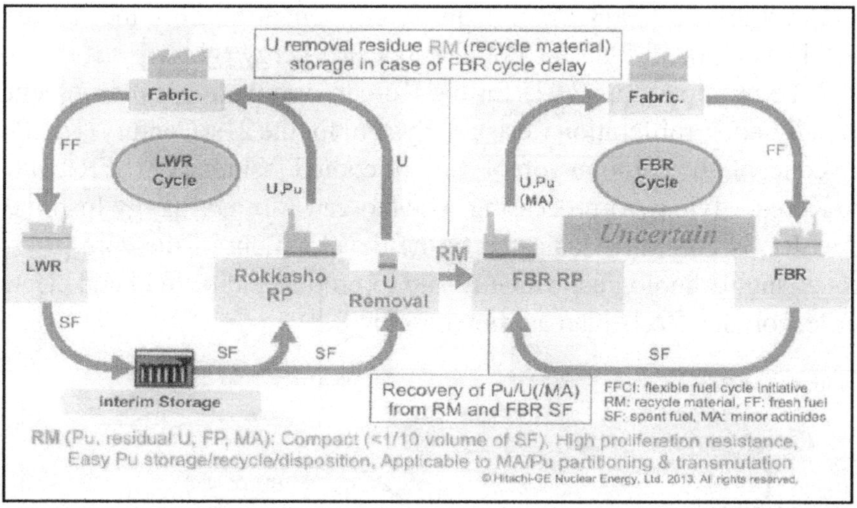

Figure 6. Proposed Flexible Fuel Cycle Initiative System in Japan[39]

South Korea

South Korea is not permitted by the United States to pursue aqueous reprocessing process and has to develop a dry reprocessing process, such as the Direct Use of PWR spent fuel In Canada deuterium uranium (CANDU) reactor (DUPIC) process developed in the 1990s. Although the DUPIC program was stopped, the expertise accumulated in the R&D of the DUPIC program laid the foundation of the later pyro-processing program because DUPIC itself was a pyro-processing technology.[40]

---

39. Ibid.

40. I. S. Hwang and K. C. Song, "Status and plans for nuclear energy and fuel cycle in the Republic of Korea," presented at the 5th Annual Workshop on Nuclear Energy Nonproliferation in East Asia, June 28- July 1, 2009, Busan, Republic of Korea and K.C. Song, "Sustainable nuclear energy and spent fuel management in Korea,"presented at the "Workshop on Multilateral Concepts of Nuclear Fuel Cycle in Asia Pacific," Tokyo, Japan, October 26-27, 2009.

On December 22, 2008, the Korean government announced the "Long-term Development Plan for Future Nuclear Energy System." It stipulated the promotion of R&D for the Korean, Innovative, Environment Friendly, and Proliferation Resistant System for the 21st Century (KIEP-21) concept, the combination of sodium cooled fast reactor (SFR) with the Korean style pyro-processing. The Korean Atomic Energy Institute (KAERI) is currently dedicated to the development of the core pyro-processing technologies, which is said to produce "dirty fuel and clean waste," on this R&D plan as shown in Fig 7.[41]

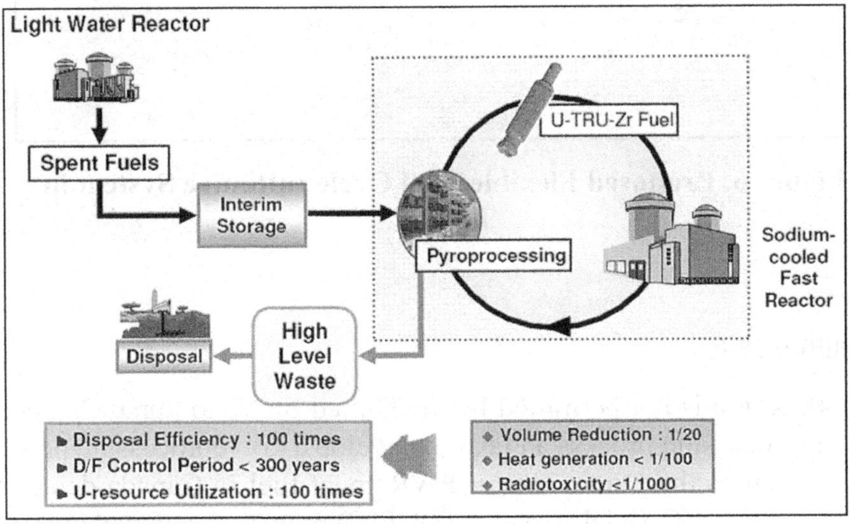

Figure 7. Schematic of Korea's R&D on SFR and Pyro-processing Technologies.[42]

- According to South Korea's national plan, "Conceptual Design & Component Tests" of SFR will be completed by 2016. The "Demonstration Reactor Design & Construction" will be carried out in 2017-2030. For the pyro-processing, the lab-scale studies have been completed by 2009. The engineering-scale verification will be completed

41. Jerng.

42. Ibid.

by 2016. The active demonstration test will be carried out in 2017-2025. The demonstration operation and "TRU Fuel Supply" will be started in 2026.[43]

- South Korea's long-term plan for SFR and pyro-processing, with the aim for achieving SFR recycle by 2030, is shown in Figure 8.

Key pyro-processing technologies have been developed by KAERI and an inactive engineering–scale mock-up facility (10 t/a), named Pyro-process Integrated Inactive Demonstration (PRIDE) facility, was constructed in 2012, aiming at an engineering–scale verification to judge its economic benefits and full technical feasibility as well as nonproliferation aspects.[44]

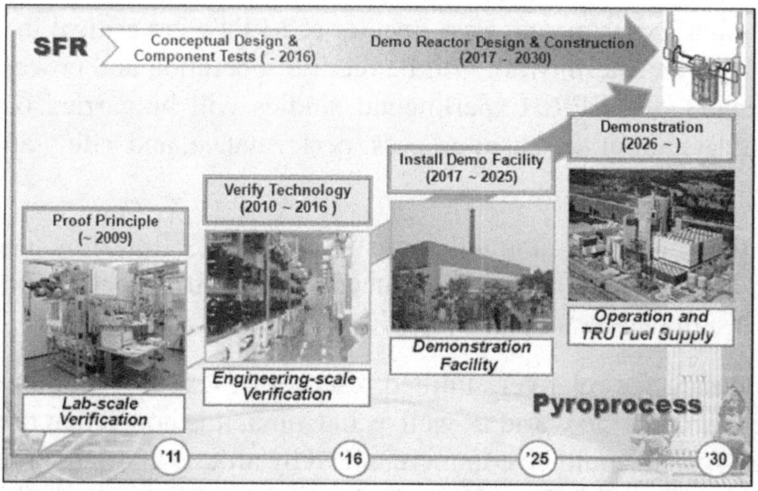

Figure 8. Korea's Long-term Plan for SFR and Pyro-processing.[45]

43. Hwang and Song; and Chan Bock Lee, Byoung Oon Lee, Ki Hwan Kim, and Sung Ho Kim (KAERI) "Status of SFR Metal Fuel Development,"Presentation at the International Conference on Fast Reactors and Related Fuel Cycles, Paris, France, March 5, 2013, p. 17.

44. Ko and Lee.

45. Hwang and Song.

China

The sustainable development of nuclear energy in China should include: (1) The large scale deployment of third generation PWRs in the coming 30 years or so; and (2) the smooth transition of nuclear energy from third to fourth generation. To support the ambitious nuclear energy program, China has to build an integrated fuel cycle industry. No one else in the world can do such a big job for such a big nuclear program in China.

With the introduction of FRs and the closed fuel cycle, it is possible to multiply by 50-60 times the energy produced from a given amount of uranium and to reduce significantly (by a factor of 10) the amount of highly radioactive waste for geologic disposal.

The China Experimental Fast Reactor (CEFR) went critical in July 2010. The near-term work will be the trial operation and power enhancement of CEFR. Experimental studies will be carried out to obtain data about the basic process, performance, and safety and to accumulate experience.

With the aim of building a demonstration FR (CFR600) by around 2023, a great deal of R&D work needs to be done to get the necessary design parameters based on the CEFR.

Considering the relatively limited reserves of low cost uranium resources around the world as well as in China, it is hoped that the FR energy system could be commercialized by around 2040. In order to reach this goal, R&D work on CFC must be carried out simultaneously. The FR fuel cycle includes PWR spent fuel reprocessing, FR fuel fabrication, FR spent fuel reprocessing, high level waste treatment, and disposal, as shown in Figure 9.

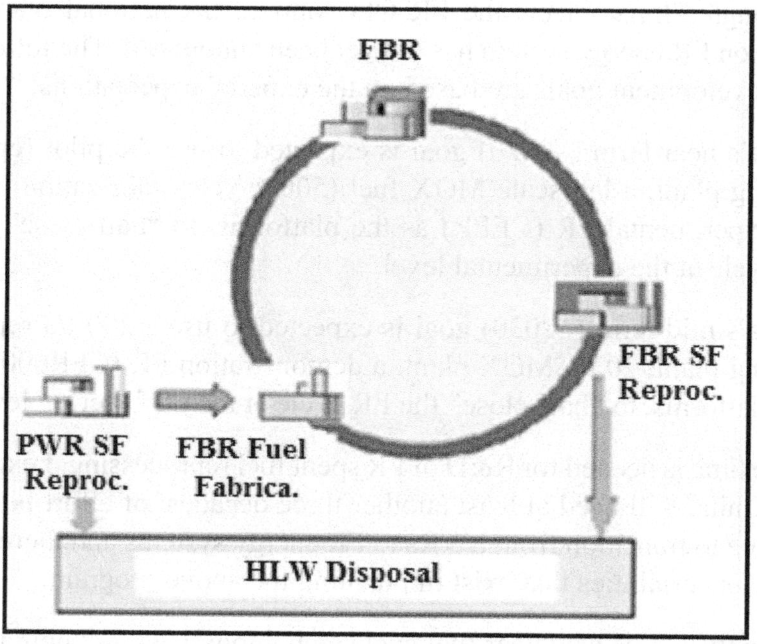

Figure 9. Schematic of FR Fuel Cycle

China's PWR spent fuel reprocessing pilot plant, with a task of processing 50 t spent fuel, completed hot testing in December 2010. A commercial reprocessing plant is under consideration and is expected to be completed by around 2020. The large reprocessing plant to be built in China will be designed based on the experience of pilot plant operations and the experience gained in developed countries. International co-operation will be very important for designing and building the commercial reprocessing plant in China.

The development of MOX fuel fabrication is now at its early stage in China. It is expected that the MOX fuel pellet could be fabricated for irradiation tests by 2015 and the MOX fuel assembly could be fabricated for irradiation tests by 2018. The MOX fuel will be first used in CEFR to get some experience. MOX fuel will be eventually replaced by U-Pu-zirconium (Zr) metal alloy fuel so as to improve the breeding performance of the fuel in FRs. Therefore, this metal alloy fuel should also be developed in the coming years.

Although China selects the FR CFC option, the national plan of R&D on FR energy system has not yet been announced. The following development goals are based on the experts' expectations.

China's near-term (~2020) goal is expected to use the pilot reprocessing plant, a lab-scale MOX fuel (500kg/year) fabrication line, the experimental FR (CEFR) as the platforms, to "half-close" the FR cycle at the experimental level.

China's mid-term (~2030) goal is expected to use a 200 t/a reprocessing plant, 20 t/a MOX plant, a demonstration FR (CFR600) as the platforms, to "half-close" the FR cycle at the engineering level.

More time is needed for R&D of FR spent fuel reprocessing. I expect that China will need at least another three decades' of effort before starting to transition from LWRs to FR energy systems and there are many uncertainties that exist in pursuing the above program.

Presently, China's nuclear fuel cycle technology lags far behind the world's advanced level. Up to now, China has not had the industrial capability in the back-end of the fuel cycle.

To achieve the above mentioned ambitious goal, the Chinese government must pay special attention to the FR energy system program and organize it carefully in a unified way. A development roadmap for China's FR nuclear energy system, including the closed fuel cycle, needs to be worked out as soon as possible in order to better implement the program.

Associated 3S Issues Related to Nuclear Energy Programs in East Asia

Safety Considerations

After the Fukushima nuclear accident, safety has been the top priority for all countries with nuclear energy programs. Safety criteria, design, and regulation have been greatly strengthened worldwide, including in East Asian countries.

The world nuclear community has reached a consensus that crisis management for a major nuclear accident is beyond a single country's capability and environmental pollution often spreads across borders. International and regional cooperation and coordination can raise the safety level of nuclear energy effectively.

Safeguards/Nonproliferation Considerations

Generally speaking, all three East Asian countries, Japan, South Korea, and China, have shown good records of implementing the IAEA safeguards.

Japan, South Korea, and China are all signatory states of the Nuclear Nonproliferation Treaty (NPT) and are obliged to obey the NPT's nonproliferation standards. According to the NPT, all Japanese and Korean nuclear facilities are under strict IAEA safeguards. As a nuclear-weapon state, some of China's civil nuclear facilities are also IAEA safeguarded according to the Voluntary Offer Safeguards Agreements in force with the five nuclear-weapon states.

According to China's present back-end fuel cycle capability, China will not have a significant amount of separated plutonium before 2030. The separated plutonium will be used to fuel the FR energy system in the future. As a nuclear-weapon state, China will not use civil plutonium to make explosive devices.

As is described in the above section on South Korea, because the United States does not allow South Korea to pursue aqueous reprocessing, the country has developed pyro-processing technology and claimed that this process does not separate plutonium completely from other trans-uranium elements. Worrying about possible proliferation risks, U.S. experts argued that the final separation of pure plutonium would be relatively trivial and make plutonium much more accessible.[46] The 1974 U.S.-Korea nuclear cooperation agreement expired in 2014, and the two sides completed another 40-year agreement in June 2015. It is very difficult for the United States to agree to let Korea pursue a pyro-processing program because of the possible nonproliferation implications. However, South Korea now has the problem of having limited space in which to store the ever increasing amount of spent fuel, which will restrict the further expansion of nuclear power. It is likely that the Koreans will strongly ask for permission from the United States to pursue a pyro-processing program, similar to Japan's efforts to get the right to reprocess in the 1970s.

Some U.S. experts worry that even if the 10-year U.S.-Korea study (due for completion in 2021) was to conclude that this technology is economically feasible and offers adequate proliferation resistance, Seoul could not build a commercial size pyroprocessing plant for at least two decades. Hence, even if the United States were to consent to pyroprocessing in the text of the new agreement, it would not immediately solve South Korea's urgent spent fuel problem.[47]

One thing that must be noted is the build-up of Japan's separated plutonium, which is of special concern for both nonproliferation and

---

46. Frank van Hippel, "South Korea Reprocessing: An Unnecessary Threat to the Nonproliferation Regime," *Arms Control Today* 40, no. 3, March 2010.

47. Fred McGoldrick and Duyeon Kim, "Decision Time: U.S.-South Korea Peaceful Nuclear Cooperation," in Nicholas Hamisevicz, ed., *On Korea: Academic Paper Series* 7, Washington, DC: Korea Economic Institute of America, 2014, pp. 79-102, available from *http://www.keia.org/sites/default/files/publications/aps_march_2013_mcgoldrick-kim_-_final.pdf.*

security. In East Asia, Japan is now the only country that owns civil separated plutonium. Table 1 shows Japan's inventory of separated plutonium from 1998 to 2012. From this table, one can see that Japan has a large inventory of separated plutonium. While there has been some reduction of separated plutonium stored in France and the United Kingdom since 2008 owing to MOX fuel use, the separated plutonium stored in Japan has been steadily increasing over the past decade.

Year	Stored in Japan t	Stored in Europe t	Total t
1998	4.9	24.4	29.3
2004	5.7	37.4	43.1
2008	8.7	38.0	46.7
2012	9.3	34.9	44.2

Table 1. Japan's Inventory of Separated Plutonium from 1998 to 2012.[48]

It must be pointed out that Japan has promised to implement its "no plutonium surplus policy" since 1991, which means that plutonium produced from reprocessing is balanced by its consumption in reactors so as to avoid the possible transfer of plutonium for weapon purpose. And, in August 2003, the Japan Atomic Energy Commission (JAEC) announced its new guidelines for plutonium

48. 1998 Figures: Jor-Shan Choi, "Separated Civil Plutonium Inventories: A Comparison Between Infcirc 549 and Calculations Using Open Sources," presented at the IAEA Technical Committee Meeting on "Perspective of Utilization and Disposition of Plutonium," Brussels, Belgium, October 5-7, 2000. 2004 Figures: Tadahiro Katsuta and Tatsujiro Suzuki, "Japan's Spent Fuel and Plutonium Management Challenges," *Research Report*, no. 2, Princeton, NJ, International Panel on Fissile Materials, September 2006. 2008 Figures: Tatsujiro Suzuki, "Civilian Nuclear Power and Nuclear Non-Proliferation: How to Minimize Risks Associated with Expansion of Nuclear Fuel Cycle Activities?" presented at the University of Tokyo workshop on "Multilateral Concepts of Nuclear Fuel Cycle in Asia Pacific," Tokyo, Japan, October 26 -27, 2009. 2012 Figures: Suzuki, "Nuclear Energy and Nuclear Fuel Cycle Policy Options After the Fukushima Accident."

management preparing for commissioning of the first commercial reprocessing plant. Utilities are expected to submit plutonium usage plans annually before separation of plutonium begins.[49]

However, the inventory of separated plutonium of Japan indicated in Table 1 shows a large surplus of plutonium, which is against the "no plutonium surplus" promise. The world community expects Japan to take concrete measures to fulfill the above promise. It is reasonable that production and consumption of the separated plutonium in Japan should be balanced, and thus, the start-up of the Rokkasho reprocessing plant should be delayed for some time to achieve a reduction of the plutonium surplus in the coming years.

The internationalization of the fuel cycle is another important topic. The non-discriminatory and effective access to the peaceful use of nuclear energy is the legitimate right of all countries equally granted by the NPT. Meanwhile, the international nonproliferation regime should be enhanced so as to lower the risk of nuclear proliferation for securing the world.

With the global nuclear renaissance that has occurred since the beginning of the 21st century, quite a few emerging countries are planning or have announced plans to develop nuclear power.

However, the peaceful use of nuclear energy is always accompanied by the possibility of nuclear proliferation. This so-called "nuclear dilemma" has been nagging the world community for more than half century. With the nuclear renaissance, this "nuclear dilemma" becomes an ever increasing concern of the world community.

The concept of "International Fuel Cycle Centers," which is an old idea, has been brought up again by the IAEA, the United States, Russia, and other countries since the early 2000s, with the aim to promote nuclear energy development while lowering the risks of nuclear proliferation. The key point of the internationalization of

---

49. Suzuki, "Nuclear Energy and Nuclear Fuel Cycle Policy Options After the Fukushima Accident."

the fuel cycle is to restrict fuel cycle activities to a limited number of countries.

The concept of internationalizing the nuclear fuel cycle seems good but has many implications in terms of technical feasibility, economic competitiveness, political issues, and public acceptance. These problems need to be solved gradually through extensive international dialogue, maybe first at the experts' level. It is reasonable to promote the internationalization of the fuel cycle step-by-step. It would be easier with the front end of the fuel cycle, but much more difficult with the back end.

In East Asia, Japan, South Korea, and China all have CFC programs. To achieve regional cooperation on the nuclear fuel cycle, mutual trust is the most important thing; yet, presently the political basis of mutual trust in East Asia is lacking. The United States has great influence in this region as well as in the world. It is the author's hope that the United States could make more contributions with an objective and fair attitude in promoting the cooperation in this area in the region.

Security Considerations

With the rapid development of nuclear power in East Asia, the accumulation of ever increasing spent fuel is of great concern.

All the East Asian countries (Japan, South Korea, and China) have reprocessing/recycling plans. However, because of the limited reprocessing capacity envisioned, the amount of spent fuel reprocessed will be a small portion of what will be generated. It is expected that there will be large accumulations of spent fuel in this region in the coming decades as described above.

It is reported that many of South Korea's NPPs will likely reach their capacity for storing spent fuel in their pools by 2020 and some of the pools will be full by 2016. However, the Korean government has yet to designate additional capacity that would ensure continued operation of the NPPs.[50] South Korea is trying to implement the pyro-processing-SFR program for recycling uranium resources and reducing the build-up of spent fuel. But the technical and economic feasibility of this program still must be proved and it is doubtful that Korea will be able to commercialize its pyroprocessing-SFR program before 2030. Japan is also confronted with the challenges of building additional storage capacities of spent fuel after re-start of its NPPs in 2014. Although China has decided to build a reprocessing plant by 2020 or so, the reprocessing capability is too small compared to the spent fuel generated. Therefore, China is planning to build some additional pools or dry storage facilities for spent fuel. In any case, establishing greater storage capabilities for spent fuel in these countries is inevitable in the coming decades.

So far, it can be said that all the spent fuel stored in East Asian countries are under safe and secure management. But, with the further increased build-up of spent fuel in this region, there exist potential risks of nuclear terrorism. The NPPs and the fuel cycle facilities with nuclear materials (e.g. HEU and separated plutonium) are targets that could be attacked by terrorists. In the past decades, nuclear terrorism has become a big threat to nuclear security. IAEA data shows that during 1993-2006, of the 1,080 intercepted illicit trafficking events, some 25% of cases involved nuclear materials (See Figure 10).[51]

---

50. Ferenc Dalnoki-Veress, Miles Pomper, et al., "The Bigger Picture: Rethinking Spent Fuel Management in South Korea," *Occasional Paper,* no. 16, Monterey, CA: James Martin Center for Nonproliferation Studies, February 2013.

51. Michael May, et al.,"Nuclear Forensics: Role, State of the Art, Program Needs," Washington, DC: Joint Working Group of the American Physical Society and the American Association for the Advancement of Science, November 2008, pp. 15-16, available from *www.aps.org/policy/reports/popa-reports/upload/nuclear-forensics.pdf.*

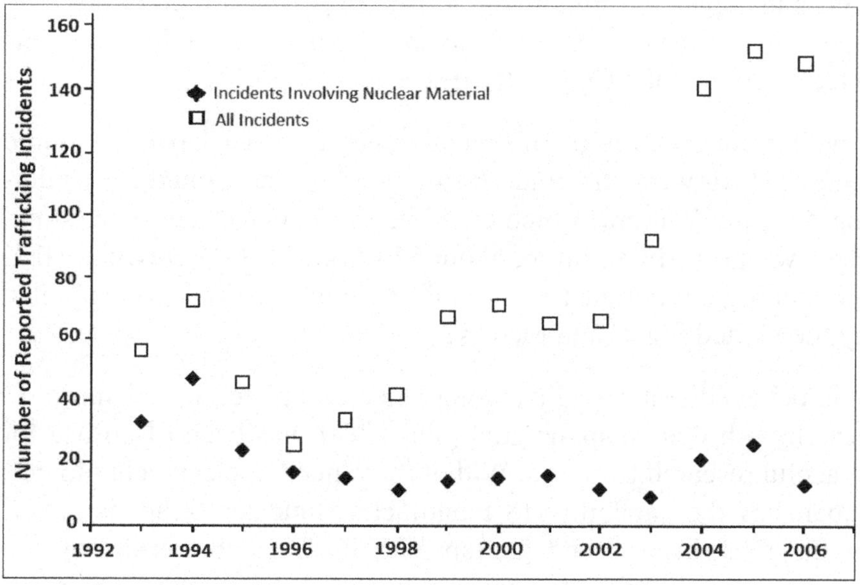

Figure 10. IAEA's Incident and Trafficking Database (1993-2006) for Illicit Trafficking Events.[52]

To combat nuclear terrorism effectively, physical protection of nuclear facilities and nuclear materials (especially separated Pu) must be strengthened. Each country has the responsibility to secure its nuclear facilities and nuclear materials. Over the past decades, China has very good records in securing its nuclear facilities and nuclear materials.

As shown in Figure 10, nuclear terrorist activities are often across borders. So, international and regional cooperation is of vital importance in combating nuclear terrorism. For example, we need to share the necessary information in the area of nuclear forensics.

A good example of international cooperation in combating nuclear terrorism is the China-U.S. joint Center of Excellence (COE) under construction in a southeast suburb of Beijing. There are similar

52. International Atomic Energy Agency, "Incident and Trafficking Database (ITDB)," last updated December 9, 2014, available from *www-ns.iaea.org/security/itdb.htm*.

COEs in Japan and South Korea. It is important to have coordination between these COEs so as to share experience and increase the effectiveness of the COE activities.

Another important issue of special concern is that Japan has stored some 331 kg weapons-grade Pu supplied by the United States during the Cold War era, which could be used to produce some 60 nuclear weapons. In addition, about 214.5 kg of 93%-enriched HEU (containing a reported 199.5 kg of uranium-235) was also supplied by the United States and the UK.

It is believed that Japan has long since completed its scientific research with this weapons-grade Pu, which has been of no use for peaceful research for years. With its advanced nuclear technologies, Japan has the capability to manufacture nuclear weapons "overnight." It has been argued that Japan has hidden ambitions to develop nuclear weapons under the cover of peaceful use of nuclear energy.[53] Japan's Prime Minister Shinzo Abe once said that Japan's constitution did not necessarily ban possession of nuclear weapons.[54] In fact, Japanese right-wing politicians have repeatedly clamored that Japan should own nuclear weapons, which would be a serious threat to the region and the world.[55]

The international community must closely watch such development and keep on high alert.

---

53. Dennis Lee, "A Nuclear Japan: The Push for Weaponization,"*Harvard International Review* 35, no.1, Summer 2013.

54. Johnathan Schell, *The Seventh Decade: The New Shape of Nuclear Danger*, New York: Macmillan, 2007, p. 145.

55. Hiroshi Hiyama, "Ishihara calls for nuclear-armed Japan," *Japan Today*, November 21, 2012, available from *www.japantoday.com/category/politics/view/ishihara-calls-for-nuclear-armed-japan*.

Integrated Consideration of 3S Issues

The use of nuclear power and materials has inherent safety, security, and safeguards (3S) concerns. Recent concern over nuclear proliferation and terrorist attacks led to an increased emphasis on safeguards and security issues as well as the more traditional safety considerations. The integrated 3S concept was first proposed by experts at the U.S. Sandia National Laboratory (SNL) in 2007.[56] It seems reasonable that the integration of the 3Ss could decrease cost and increase effectiveness, especially in designing a new nuclear facility.

In practice, the 3Ss are often linked together. One of the examples is the process of taking away damaged nuclear fuel from Fukushima No. 1 NPP.[57] Because some of the damaged nuclear fuel rods may be melted and deformed, more advanced machines and tools should be used to ensure the safety of the workers and the environment when removing the highly radioactive and contaminated fuel rods. Meanwhile, these fuel rods are nuclear materials and concerns for safeguards/nonproliferation also arise for Japan in the eventual recovery of the core debris and nuclear materials from Fukushima. Being a non-nuclear-weapons state it would be an unprecedented challenge for Japan to account for all the nuclear materials in the reactor core as stipulated in its comprehensive safeguards agreement with the IAEA since the core materials may have leaked to outside the reactor vessels. It can be understood that such accounting for nuclear materials was not required previously in the Three Mile Island and Chernobyl accidents, as both the United States and the Soviet Union were nuclear-weapons states for whom safeguards

56. John Darby, Karl Horak, et al., "Framework for Integrating Safety, Operations, Security and Safeguards in the Design and Operation of Nuclear Facilities," Sandia Report SAND 2007-2469, Albuquerque, NM: Sandia National Laboratory, October 2007.

57. Gu Zhongmao, "Re-thinking of Nuclear Energy and the 3S Issues After Fukushima Nuclear Accident," presented at the International Institute for Strategic Studies-China Institute of International Studies"Workshop on Nuclear Security," Qingdao, China, June 14-15, 2012.

agreements with the IAEA were voluntary.[58]

In addition, nuclear security problems will exist during the decades-long process of removing nuclear materials from the pools and reactor vessels of Fukushima No.1 NPP. To lower the risks of a terrorist attack, nuclear security measures must be taken to prevent, detect, and respond to the theft, sabotage, unauthorized access, illegal transfer, or other malicious acts involving nuclear materials or the associated facilities.

58. Jor-Shan Choi, "Evaluating the 3S (Safety, Security, and Safeguards) Aspects of Advanced Reprocessing Technologies in Nuclear Fuel Cycles and Radioactive Waste Management," presented at the 2nd China-Japan "Academic Symposium on Nuclear Fuel Cycle," Shanghai, China, November 27-30, 2013.

Appendix

Meeting China's Future Energy Needs and Environmental Commitments: Is Increasing Nuclear Power the Best Way to Go?

David Von Hippel

Introduction: China's Energy and Environmental Challenges

Projections of Economic Development and Energy Use in China, and Their Environmental Price

China's emergence as a global economic powerhouse from what was a largely rural and agrarian society even as of the 1980s has been a defining characteristic of the past few decades. The rate of growth in gross domestic product (GDP) in China exceeded 7% per annum in each year between 1991 and 2015, dipping only slightly below 7% in 2015 and 2016.[1] The result of this rapid growth has been that China's economy is more than 11 times larger today than it was in 1990, and nearly 30 times larger than it was in 1980. An increase in energy production and use, and especially electricity generation, has fueled the growth in China's economy. Primary energy use, a measure of all of the fuels that go into providing energy for an economy, has increased by nearly a factor of five since 1990, as has oil and oil products consumption, and electricity generation

1. World Bank, "GDP Growth, Annual (%)," graph and data, accessed 2017, available from *http://data.worldbank.org/indicator/NY.GDP.MKTP.KD.ZG?locations=CN.*

has grown nearly 10-fold in the same period. China edged the United States as the world's largest producer of coal in the mid-1980s. In 1990 coal production in China was over a billion tonnes (metric tons) per year, and rose to nearly 4 billion tonnes by 2013, before declining somewhat in recent years.

Although it has significant energy resources of its own, China has been obliged to turn to imports to help fuel its massive economic growth. China became a net importer of oil and oil products in the early 1990s, and a net importer of gas, as mostly as liquefied natural gas (LNG), in about 2006. By 2016, China was the world's third-largest importer of LNG, after Japan and the Republic of Korea (ROK). China's oil imports broke a monthly record early in 2017, with its imports exceeding those of the United States to be the world's highest.[2] China sources its oil imports from a diverse array of nations; by 2014, 14 countries each supplied at least 2% of China's total oil imports, with Saudi Arabia its leading supplier at 16%.[3] China's energy imports dependency continues to increase, as consumption continues to rise while production of coal and oil, in particular, are relatively static. China's energy imports dependency is not yet, however, at the 90-plus percent level found in the ROK and Japan.

The large and rapid increase in China's energy use has been accompanied by a variety of environmental challenges. In 2006, China passed the United States to become the largest emitter of greenhouse gases (GHGs) among nations,[4] and by 2014 emitted 30% of global anthropogenic GHG emissions, although its per capita emissions re-

2. See, for example, Gordon Kristopher, "China's Crude Oil Imports Hit a New Record," *Market Realist*, April 19, 2017, available from *http://marketrealist.com/2017/04/chinas-crude-oil-imports-hit-new-record/*.

3. United States Department of Energy, "Energy Information Administration, China," last updated May 14, 2015, available from *https://energy.gov/sites/prod/files/2016/04/f30/China_International_Analysis_US.pdf*.

4. See, for example, Carbon Brief, "Global Historical Emissions Map," available from *https://www.carbonbrief.org/interactive-map-historical-emissions-around-the-world*.

main less than those in the United States and many other industrialized nations.[5] In addition, increased energy use in China has led to emissions of local and regional air pollutants high enough to be a significant danger to health much of the year in many Chinese cities, including Beijing,[6] and has contributed to water pollution, soil degradation, and numerous other environmental problems.

Although China has markedly improved energy use efficiency in recent years, particularly as measured in energy use per unit of GDP, shifting away from coal use to reduce global, regional, and local environmental problems have continued to be a priority in Chinese policy. This shift is to be in part accomplished by moving some of the most polluting industries out of the major southern and eastern Chinese cities to other areas of China, as well as moving some of those industries to other nations as the Chinese economy moves more toward high value-added manufacturing and services. In the electricity sector, this means a combination of improved efficiency in existing and new coal fired power plants and increasing the share of power generated from non-fossil resources, including renewable energy sources (particularly hydroelectric, wind, and solar power) and nuclear power.

China's nuclear sector is young by comparison to that of Japan and the ROK, but is and has been growing fast, as most of the reactors built worldwide at present are being built in China. With a large land area and a not-yet-powerful civil society sector, siting of nuclear plants and spent fuel facilities has not yet been a major problem for China, though it may grow to be so in the future.

---

5. United States Environmental Protection Agency, "Global Greenhouse Gas Emissions Data," 2016, available from *https://www.epa.gov/ghgemissions/global-greenhouse-gas-emissions-data*.

6. See, for example, "China orders cities to reduce emissions, as capital braces for another bout of heavy smog," *South China Morning Post,* February 13, 2017, available from *http://www.scmp.com/news/china/policies-politics/article/2070392/china-orders-cities-reduce-emissions-capital-braces*.

Past and Projected Patterns of Growth of the Chinese Electricity Sector

Total Electricity generation in China in 1990 was approximately 650 terawatt hours (TWh, or billion kilowatt-hours), including Hong Kong, which is about the same as present-day Germany, with about 7% of China's population.[7] By 2016, electricity output and consumption in China had grown nearly 10-fold, supplanting the United States in 2011 as the nation with the largest electricity consumption (see Figure 1). Overall generation capacity grew even more rapidly, particularly in recent years, with growth in capacity averaging over 9% annually from 1990 through 2005, and 11% annually from 2005 through 2016 (see Figure 2).[8] Generation capacity in China now exceeds 1600 GW (gigawatts, or million kilowatts), nearly 60% more than the United States, where generation capacity stood at a bit over 1000 GW as of 2016.[9]

Thermal power, and specifically coal-fired power, has been the mainstay of Chinese electricity generation. Thermal power provided about 80% of generation in 1990, remaining near that level though 2010, falling only in recent years to under 72% by 2016. Despite rapid growth in nuclear capacity, nuclear power accounted for only 3.6% of electricity output by 2016, somewhat less than wind power in that year. Construction of hydroelectric capacity has been rapid in the past decade and continues today with, nearly 12 Gigawatt electrical (GWe) of capacity added in 2016 alone to a total of 330

---

7. British Petroleum, *BP Statistical Review of World Energy*, June 2017, excel workbook accompanying volume available from *http://www.bp.com/statistical-review*.

8. Some of the even more rapid growth in the post-2005 period was due to the addition of large amounts of wind and solar power generation capacity, which have lower capacity factors, and thus generate less energy annually per unit of capacity than, for example, coal-fired and nuclear power plants.

9. United States Department of Energy, Energy Information Administration, "Electricity in the United States," 2017, available from *https://www.eia.gov/energyexplained/index.cfm?page=electricity_in_the_united_states#tab2*.

GWe (of which about 27 GWe are pumped-storage plants used for peak power provision).[10] Electricity consumption in China has been dominated by the industrial sector, which consumed nearly 77% of power in 1990. The importance of the industrial sector has waned somewhat—to about 70% of total consumption in 2014, as residential and commercial/services electricity use has grown—but still remains the major user of electricity in the Chinese economy.

The rapid overall growth in electricity consumption in the last decade, however, masks much slower growth in recent years—just over 3% annually between 2013 and 2016, as shown in the last four bars of Figure 2—as the Chinese economy has slowed somewhat, and greater emphases have been placed on improvements in energy efficiency, development of the services sector, and the reduction of heavy industry.

10. International Hydropower Association, "China," last updated May, 2017, available from *https://www.hydropower.org/country-profiles/china*. This reference highlights the construction of the "Wudongde project on the Jinsha River in the south- west, which will provide 10.2 GW installed capacity when complete" in 2020, and will be the sixth-largest hydro plant in the world.

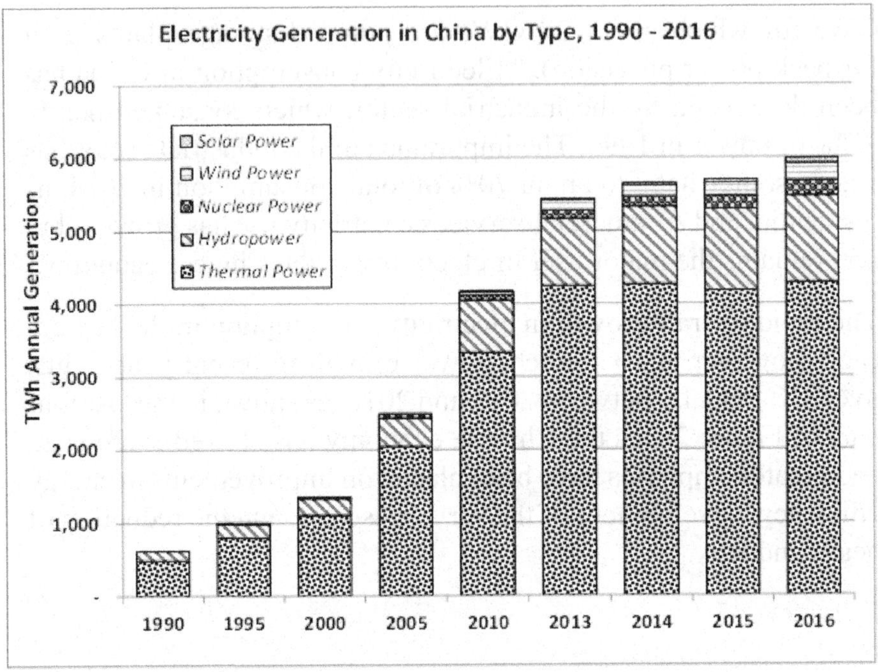

Figure 1. Electricity Generation in China by Type, 1990 through 2016.[11]

11. Sources: Figures through 2014 from *China Statistical Yearbook*, 2016, available from *http://www.stats.gov.cn/tjsj/ndsj/2016/indexeh.htm*, with 2015 and 2016 data from John A. Mathews and Hao Tan, "China's Continuing Green Shift in the Electric Power Sector: Evidence from 2016 data," *The Asia-Pacific Journal Japan Focus* 15, Issue 10, no. 4, May 15, 2017, available from *http://apjjf.org/-John-A--Mathews--Hao-Tan/5038/article.pdf*.

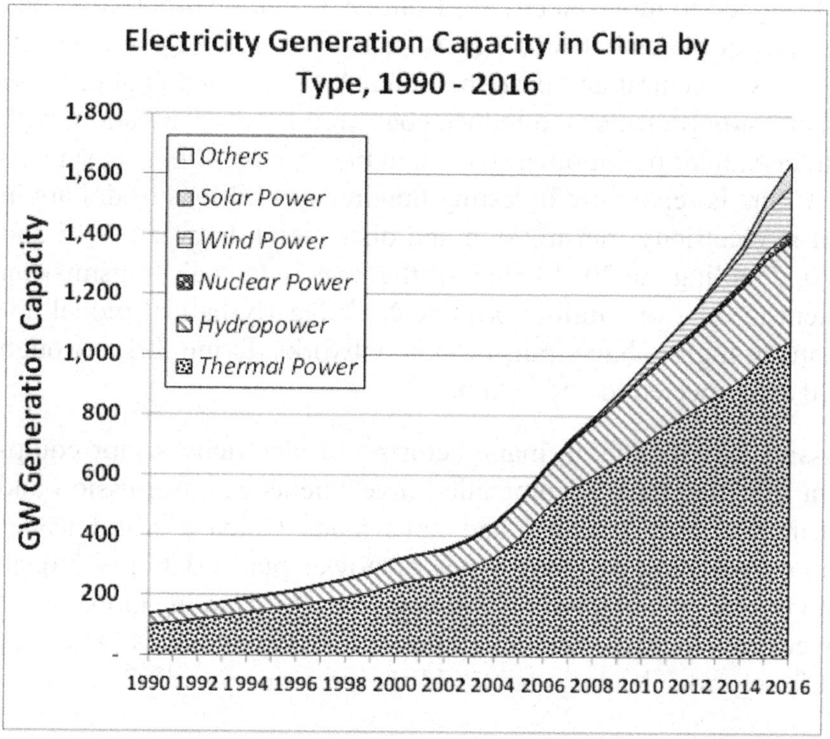

Figure 2. Electricity Generation Capacity in China by Type, 1900 through 2016.[12]

Although electricity in the 1990s and 2000s was transmitted and distributed through mostly six regional grid "clusters," recent years have seen massive investments in transmission lines designed to tie the national grid together. In addition to several point-to-point UHV DC (Ultra-high voltage Direct Current) lines, China has invested and is investing tens of billions of dollars in UHV AC (alternating current) lines.[13] Together, these UHV DC and UHV AC lines

12. Sources of data are as indicated for Figure 1.

13. UHV DC lines are 800 kV (kilovolts) or more, and UHV AC lines are 1000 kV or more. By way of comparison, large high-voltage transmission lines in most countries are rated at about 500 kV. The higher the voltage, the more power can be carried by a given line. See, for example, "China Exclusive: China to

are designed to move electricity from coal-fired, hydroelectric, and wind power plants in China's North and West to the major consuming cities in central and eastern China. An additional goal of these transmission projects is to reduce coal-fired power generation, with its attendant air pollution problems, in the vicinity of big coastal cities. China is reportedly investing hundreds of billions of dollars in total in electricity transmission and distribution between 2015 and 2020, doubling the 2014 length of the lines in China's transmission system to over one million kilometers.[14] Nearly half of global additions to high-voltage transmission networks during 2014 through 2020 are expected to be in China.

Massive investments in manufacturing of electricity sector equipment in China have accompanied investments in transmission and distribution infrastructure, and have made China a world leader in the production of many types of power plants. China's largest wind turbine manufacturer, Goldwing, was third among global wind power firms in 2016, with an output of 6.4 GW (almost all installed in China), after leading the world the previous year.[15] China's so-

---

build 12 power transmission lines," *China Daily*, May 14, 2014, available from *http://shanxi.chinadaily.com.cn/2014-05/14/content_17505983.htm*.

14. See, for example, "UPDATE 1-China targets $300 bln power grid spend over 2015-20 – report," *Reuters*, August 31, 2015, available from *http://www.reuters.com/article/china-power-transmission-idUSL4N1171UP20150901* and "Global Transmission and Distribution Market 2015 Report Says China Leading Power Transmission Line Additions to 2020," *Cision PR Newswire,* June 11, 2015, available from *http://www.prnewswire.com/news-releases/global-transmission-and-distribution-market-2015-report-says-china-leading-power-transmission-line-additions-to-2020-506906491.html*. By way of comparison, the U.S. transmission grid includes about 700 thousand kilometers of high voltage lines. See U.S. Department of Energy, "Top 9 Things You Didn't Know About America's Power Grid," November 20, 2014, available from *https://energy.gov/articles/top-9-things-you-didnt-know-about-americas-power-grid*.

15. See, for example, "Vestas reclaims top spot in annual ranking of wind turbine makers," *Bloomberg New Energy Finance,* February 22, 2017, available from *https://about.bnef.com/blog/vestas-reclaims-top-spot-annual-ranking-wind-turbine-makers/*.

lar photovoltaic (PV) firms produced 71% of the world's PV modules in 2016—and most of their modules were installed in China, as well as exporting panels to the rest of the world.[16] Almost by definition, China leads the world in production of coal-fired power plants, installed both in China and, increasingly, in other nations.[17] And China has slowly transitioned its nuclear industry from plants built with mostly foreign technology to plants designed and built in-country (see below).

The emphasis on manufacturing of renewable energy equipment has accompanied aggressive national goals for renewable energy deployment. Progress toward, and even past, these goals has been impressive. Deployment of solar photovoltaic power has been so rapid—topping 10 GW in a single recent month—that China's goal for solar deployment by 2020 under its 13th Five-Year Plan (FYP) has been nearly doubled, from 105 GW (already achieved) to 230 GW.[18] Progress toward wind power deployment goals has been nearly as impressive, with 129 GW of wind power capacity deployed by 2015, already over half of the 2020 target of 210 GW

---

16. Jeffrey Ball, Dan Reicher, Xiaojing Sun, and Caitlin Pollock, *The New Solar System: China's Evolving Solar Industry And Its Implications for Competitive Solar Power In the United States and the World*, Stanford University Steyer-Taylor Center for Energy Policy and Finance, March 2017, available from *https://www-cdn.law.stanford.edu/wp-content/uploads/2017/03/2017-03-20-Stanford-China-Report.pdf*.

17. See, for example, Hiroko Tabuchi, "As Beijing Joins Climate Fight, Chinese Companies Build Coal Plants," *The New York Times*, July 1, 2017, available from *https://www.nytimes.com/2017/07/01/climate/china-energy-companies-coal-plants-climate-change.html?mcubz=0&_r=0*.

18. See, for example, Joshua S. Hill, "China Continues Massive Solar Installations With 10.52 GW In July, Already Exceeds 2020 Target," *Cleantechnica*, August 22, 2017, available from *https://cleantechnica.com/2017/08/22/china-continues-massive-solar-installations-10-52-gw-july-already-exceeds-2020-target/* and Mark Osborne, "China's solar boom to continue through 2020 as install targets revised," *PVTech*, July 28, 2017, available from *https://www.pv-tech.org/news/chinas-solar-boom-to-continue-through-2020-as-install-targets-revised*.

set in China's 13th FYP for energy. For some wind-rich provinces, wind and solar power already provided up to 15% of total generation by 2020.

China's renewable energy industries have not been without growing pains. Many wind generators, particularly in northern and western provinces, were built in anticipation of local electricity demand and/or construction of transmission facilities that have yet to catch up with wind power capacity, and as a result, wind energy output worth billions of dollars has been curtailed, far more than in other wind-rich areas of the world (such as Texas). The transmission line projects described above are expected to significantly lower curtailment rates in coming years, allowing wind energy from the North and West of China to displace coal-fired power for the cities of the East and South.[19]

Accompanying this drive to use more renewable energy has been a drive toward energy efficiency in multiple sectors. China's National 13th FYP includes a reduction of 15% in energy use per unit of GDP relative to the level in 2015 and a reduction of 18% in carbon dioxide emissions per unit of GDP.[20] It should be noted that energy (and CO_2) per unit of GDP are indicators dependent on several factors, most notably, the composition of industry in China, the types of products produced, and the value of those products, as well as the actual energy efficiency per unit of physical output. CO_2 per unit of GDP additionally factors in the composition of the energy sources used by an economy. As a result, a reduction in energy use (and CO_2 emissions) per unit of GDP can be accomplished by a combination of true energy efficiency improvements, greater value added in products produced, offshoring of heavy (energy intensive/polluting)

19. See, for example, Li Ying, "Blowing in the wind," *Chinadialogue*, May, 31, 2016, available from *https://www.chinadialogue.net/article/show/single/en/8965-Blowing-in-the-wind.*

20. Ma Tianjie, "China's Ambitious New Clean Energy Targets," *The Diplomat*, January 14, 2017, available from *http://thediplomat.com/2017/01/chinas-ambitious-new-clean-energy-targets/.*

industries, and a shift toward production of more services, all of which are currently in play in China.

China's 13th Five-year Plan lays out a number of goals for electricity sector development by 2020 (see Table 1). These include overall electricity consumption (given as a range), development of various types of generation, and other parameters. Along with the aggressive targets for renewable power development described above, the 13th FYP shows growth in electricity generation/consumption, as well as generation capacity, slowing markedly relative to experience over the past decade.

Although longer-term official forecasts of electricity demand were not available for this paper, the trend of declining growth in Chinese electricity generation and consumption is echoed and extended in a number of forecasts by other analysts. For example, in the U.S. Department of Energy's (U.S. DOE's) *International Energy Outlook 2016*, the growth rate of electricity use in China progressively decreases from about 3.6% annually in 2015-2020 to 1.4%/yr in 2035 through 2040 (see Figure 3). As China's population growth will, based on the United Nations' "Medium Variant" estimate, have reached its peak just before 2030,[21] continued growth in electricity consumption late in the U.S. DOE forecast means continued growth in electricity use per person. Overall, the U.S. DOE forecast calls for average annual growth in electricity use of just under 2.5% from 2012 through 2040, resulting in a doubling of 2012 electricity use in China by 2040.

21. Data downloaded from United Nations DESA, "World Population Prospects 2017," available from *https://esa.un.org/unpd/wpp/Download/Standard/Population/*.

Category	2015 Value	2020 Target	Annual average Growth Rate [or change]
Installed Capacity (100 million kilowatts)	15.3	20	5.50%
West to East [Transmission Capacity] (100 million kilowatts)	1.4	2.7	14.04%
Total electricity consumption (trillion kilowatt hours)	5.69	6.8 - 7.2	3.6 - 4.8%
Electricity accounted for the proportion of terminal energy consumption	25.80%	27%	[1.2%]
Per capita installed capacity (kW / person)	1.11	1.4	4.75%
Per capita electricity consumption (kWh / person)	4142	4860 - 5140	3.2 - 4.4%
Non-fossil energy consumption share	12.00%	15%	[3%]
Non - fossil generation share of installed capacity	35.00%	39%	[4%]
Conventional hydropower (100 million kilowatts)	2.97	3.4	2.80%
Pumped Storage hydro (10 thousand kilowatts)	2303	4000	11.70%
Nuclear power (100 million kilowatts)	0.27	0.58	16.50%
Wind power (100 million kilowatts)	1.31	2.1	9.90%
Solar power (100 million kilowatts)	0.42	1.1	21.20%
Fossil energy power generation installed proportion	65%	61%	[-4%]
Proportion of installed capacity as coal-fired power	59%	55%	[-4%]
Coal-fired generation capacity (hundred million kilowatts)	9	<11	4.10%
Gas-fired generation capacity (hundred million kilowatts)	0.66	1.1	10.80%
Average coal consumption of new coal-fired units (grams of standard coal / kWh)		300	
Average coal consumption of active coal-fired units (grams of standard coal / kWh)	318	310	[-8]
Transmission Line Loss Rate	6.64%	<6.5%	
Charging Facilities Construction	Meet requirements for charging 5 million electric cars		
Electricity from Alternative Sources (100 million kWh) [Presumably for Transport]		4500	

Table 1. Thirteenth Five-Year Plan Main Objectives of Power Industry Development.[22]

22. Rough translation of "Box 2" from [China] National Development and Re-

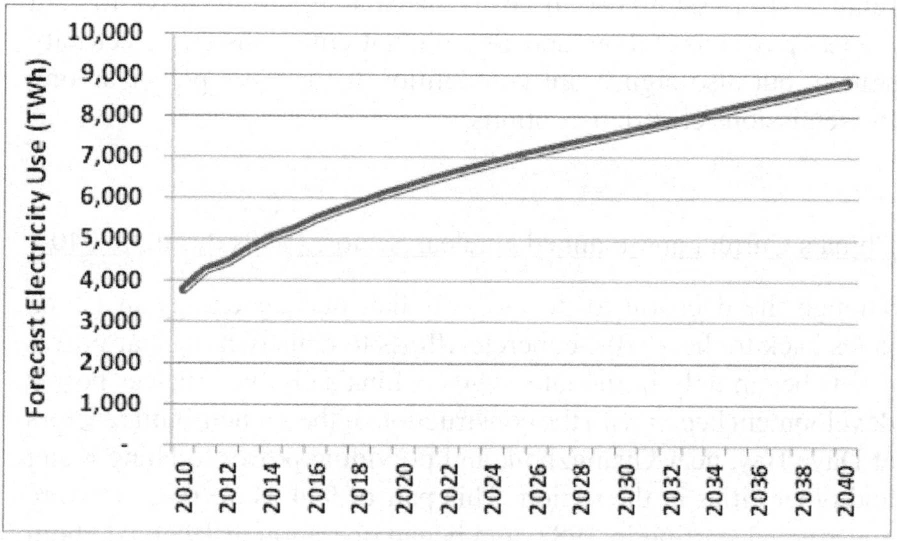

Figure 3. Forecast for Electricity Use in China, 2010-2040 (U.S. Department of Energy International Energy Outlook).[23]

The current setting for China's nuclear power sector is thus characterized by rapid (by the standards of most countries) but slowing economic growth, increasing energy efficiency, and substantial progress on many development goals. In combination, these factors are resulting in demand for electric power that will continue to grow, but at progressively lower rates as the decades pass. At the same time, an aggressive drive to use electricity from renewable sources and reduce electricity generation (and its attendant envi-

---

form Commission, National Energy Board, *The Thirteenth Five-Year Plan for Electricity Development* (2016-2020), available from *http://www.ndrc.gov.cn/zcfb/zcfbghwb/201612/P020161222570036010274.pdf*.

23. Figure prepared using data from U.S. Department of Energy's Energy Information Administration, detailed tables for International Energy Outlook 2016, "Table: Delivered energy consumption by end-use sector and fuel; Case: Reference case | Region: China," available from *https://www.eia.gov/outlooks/aeo/data/browser/#/?id=15-IEO2016&region=4-12&cases=Reference&start=2010&end=2040&f=A&linechart=Reference-d021916a.2-15-IEO2016.4-12&map=&sourcekey=0*.

ronmental emissions) from coal creates a significant drive toward nuclear power as carbon- and air pollutant-emissions-free electricity source, but also significant competition for nuclear power among low-emissions electricity options.

China's Current and Planned Nuclear Sector: 1990s through 2040

Though the decision to develop civilian nuclear energy in China dates back to the 1970s, concrete efforts to construct nuclear power plants began only in the late 1980s. China's civilian nuclear power development began with the construction of the French-built reactors at Daya Bay, near Guangzhou, and providing power to Hong Kong and other cities in the region. This pair of 944 MWe reactors were constructed starting in 1987, and began operation in 1994. At about the same time, the smaller (298 MWe) Qinshan-1 unit was built with a combination of imported and domestic technology. These three reactors constituted the "first wave" of reactor construction in China, as described by the World Nuclear Association (see Figure 4).[24] A second wave of reactor construction began in the late 1990s, and included four additional reactors at Qinshan, near Shanghai, this time in the 600 MWe class, and produced as stepping stones to development of a "Chinese standard" 1000 MWe-class unit, the CNP-1000. Two additional units at the Ling Ao power plant, very close to the original Daya Bay reactors. Like the Daya Bay plant, the Ling Ao reactors are based on French technology. A pair of reactors in northern Jiangsu province, between Shanghai and Beijing, the first set of Tianwan units, were built using Russian reactor technology but with control equipment supplied by an international consortium. These first two Tianwan units were completed as the third wave of reactor construction began.

24. Figure and related data from World Nuclear Association, "Nuclear Power in China," updated September 2017, available from *http://www.world-nuclear.org/information-library/country-profiles/countries-a-f/china-nuclear-power.aspx*.

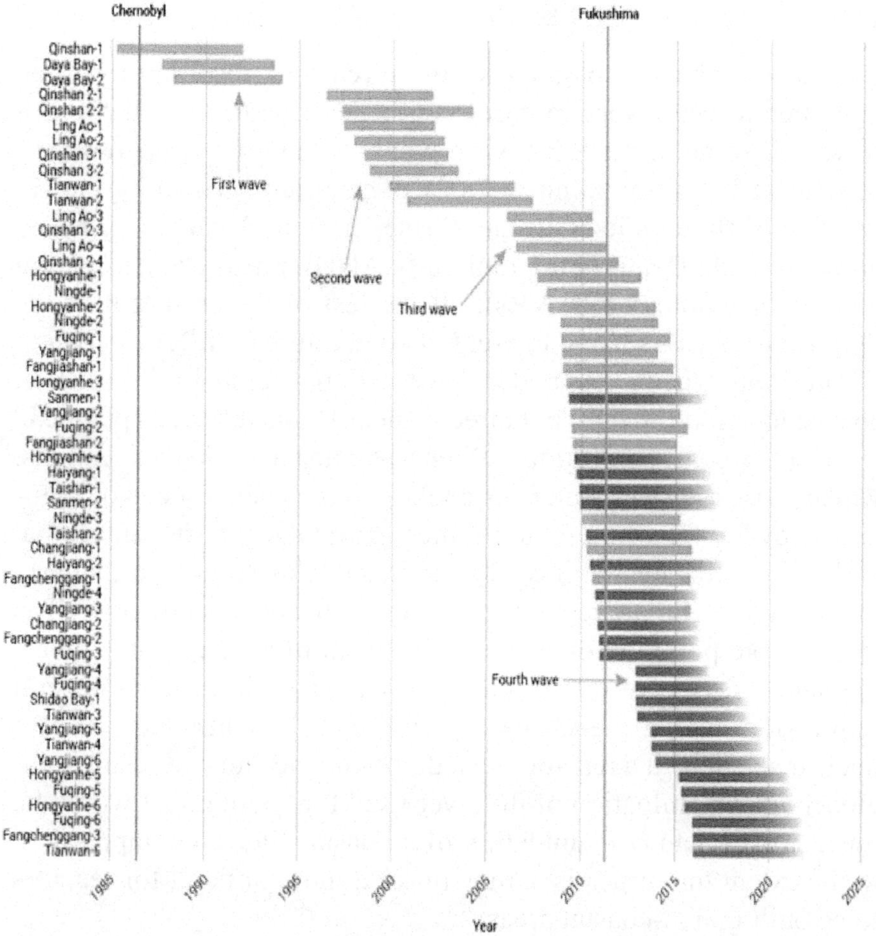

Figure 4. Nuclear Reactor Construction in China to Date.

The third wave of Chinese reactor construction is considered to have begun in about 2006, with ground broken on the last few of the 30 third wave reactors just before the Fukushima accident in Japan in March of 2011. The Chinese nuclear establishment's response to the Fukushima accident was to order a delay in new reactor starts while reactor safety provisions were reviewed and strengthened, and reactor deployment plans were reviewed. Subsequent to these reviews, a fourth wave of reactor construction is considered to have begun in 2012 and 2013 and has included at least 16 reactor units to date.

Current Fleet of Chinese Reactors

The current fleet of operating Chinese reactors, resulting from the first three waves of reactor construction described above (the fourth wave and some of the third wave plants being not yet operational) consists of 37 units totaling 33.7 GW of electricity generating capacity. Two of these units, Qinshan Phase III, units 1 and 2, are based on the Canada Deuterium Uranium (CANDU) heavy water/natural uranium technology (PHWRs); all the rest of the current fleet are light water reactors using low-enriched uranium fuel.[25] As shown in Figure 5, all existing and under-construction reactors in China are in coastal locations, and all but three nuclear plants listed as "planned" are also in coastal locations. Plants planned for inland locations, which must use river water for cooling, have been a focus of regulatory review for Chinese authorities, particularly post-Fukushima, with the reliable availability of sufficient cooling water, particularly in a changing climate regime, being a significant concern, as well as the possible pollution of rivers in the event of an accident. China's regulatory review of inland sites may have contributed to a general impression among the Chinese public that the consequences of a nuclear accident at a reactor located at an inland site will result in radioactive contamination of the river used for cooling that would be much worse than contamination of a coastal site. This impression, to the extent that it persists, may make it more difficult for reactors to be built in non-coastal areas.

25. Light water reactors, or LWRs, use regular water (H_2O) as the reactor coolant. The two primary light water reactor designs are pressurized water reactors (PWRs), which are dominant in China, and boiling water reactors (BWRs), which are also common worldwide (and of which the units at the damaged Fukushima Daiichi plant are examples). China is also building a 210 MWe high-temperature gas-cooled reactor (Model "HTR-PM") in Shandong province. The HTR-PM uses fuel spheres ("pebbles") rather than fuel encased in long metal tubes (fuel rods) like the other reactor types.

Figure 5. Location of Nuclear Power Plants Existing, Under Construction, and Planned in China.[26]

The World Nuclear Association characterizes China's nuclear policy as:

> "China has set the following points as key elements of its nuclear energy policy:
>
> - PWRs will be the mainstream but not sole reactor type.
> - Nuclear fuel assemblies are fabricated and supplied indigenously.
> - Domestic manufacturing of plant and equipment will be maximized, with self-reliance in design and project management.

26. World Nuclear Association, "Nuclear Power in China."

- International cooperation is nevertheless encouraged."

The World Nuclear Association also notes that "[t]he technology base for future reactors remains officially undefined, though two designs are currently predominant in construction plans: CAP1000 and Hualong One, after plans for more CPR-1000 units were scaled back post-Fukushima. Beyond them, high-temperature gas-cooled reactors and fast reactors appear to be the main priorities." For the present, China's use of many different kinds of reactors, ordered and funded by different provinces, and only loosely coordinated with power grid development, may prove to be problematic soon, and may complicate nationally coordinated management of spent fuel.[27]

The combination of current electricity generation over-capacity, particularly in east coast areas where many reactors are located, plus the variety of reactors under construction, the poor record of many imported reactor technologies, and China's ambitions to export reactors itself, combine to yield picture of China's nuclear future that is significantly muddled relative to stated policy. The following subsection, co-authored by Professor Stephen Thomas of the University of Greenwich, United Kingdom, briefly explores these issues.[28]

---

27. In addition, China's nuclear energy sector involves dozens of different firms and entities. Some of these are national in scope, some regional or provincial, and some organized specifically to implement a particular nuclear project. Ownership of nuclear entities is similarly complex, with national and provincial utilities being mostly state-owned, but with a variety of joint ventures and subsidiaries involving publicly traded companies, private entities, and others. See World Nuclear Association, "Government Structure and Ownership: Nuclear Power in China Appendix 1," updated September 2017, available from *http://www.world-nuclear.org/information-library/country-profiles/countries-a-f/appendices/nuclear-power-in-china-appendix-1-government-struc.aspx*.

28. For additional detail, see Steve Thomas, "China's nuclear export drive: Trojan Horse or Marshall Plan?" *Energy Policy* 101, 2017, pp. 683–691.

China's Nuclear Export Ambitions: Prospects and Challenges[29]

China's lack of focus on one or a very few reactor designs has resulted in a technologically challenging situation as a major reactor build-out continues. China's original plan as of 2007 was to use the 1970s/1980s French reactor design for a few years, but import and indigenize state-of-the-art foreign technologies, with the goal of making reactors affordable for domestic use and for export to countries new to nuclear power, as well as to established nuclear energy users. The indigenizing of foreign technologies was to have been done through the adaptation of the Westinghouse AP1000 and to a lesser extent the French (Areva) EPR designs. Both technologies have failed badly, so China's two major reactor vendors, China General Nuclear Power Group (CGN) and China National Nuclear Corporation (CNNC), have returned to the old French technology used for, for example, the units at Daya Bay, updating it and adding new features to develop the ACP-1000 and ACPR-1000 designs, which are being "merged" to yield the Hualong One. Whether this design will be safe enough to convince foreign regulators (it is being tested in the United Kingdom, which has been considering importing Chinese reactors), and still be inexpensive enough to attract buyers, remains to be seen. Another, somewhat weaker Chinese nuclear utility, SNPTC, which was set up to indigenize the AP1000 design, is scaling it up to the 1400 MWe CAP1400 design to try to make the economics more attractive, but whether doing so will be successful or not is unclear.

Despite a substantial interest in exporting reactors, and substan-

29. See also Matthew Cottee, "China's nuclear export ambitions run into friction," *Financial Times*, August 2, 2017, available from *https://www.ft.com/content/84c25750-75da-11e7-90c0-90a9d1bc9691*. Interestingly this article includes the following passage about Russia's allegedly waning interest in nuclear exports: "The current leader in the nuclear export market, Russia's Rosatom, is reportedly shifting focus to hydropower and wind turbines rather than its usual reactor business. Speaking at the 'Technoprom-2017' conference in Novosibirsk, the deputy general director of Rosatom, Vyacheslav Pershukov, suggested that the export market for nuclear reactors has been exhausted."

tial effort to do so, China has been unable to reach firm agreements except for its deal with Pakistan and an agreement, but not yet a contract, to construct (or rather, to resume construction on) two CANDU units in Romania.[30] Both of these export ventures are receiving or thought to be receiving significant Chinese financing for the purchases. A number of other deals are reportedly under discussion. The lack of additional firm contracts may be because China has only entered the field of nuclear vendors during this decade, because there is reluctance to buy from China (for reasons not completely understood) or because there is simply little actual market for reactors outside of China. The contrast between Russia, which lacks the wherewithal to provide financing and significant portions of the required the supply chain to support reactor exports, but has about 30 firm orders, and China which does have the money and supply chain but has no orders, is marked.

The size of China makes analysis of its economic and technological situations hard to carry out, because its industries can dominate international markets without being dominant at home. Nuclear represents a negligible part of China's electricity mix, yet reactors built in China account for the majority of the world's nuclear construction over the last decade. As a result, one might say that that global nuclear industry needs China, but China doesn't necessarily need nuclear power. As such, the opposition by some citizen groups to building nuclear power at inland sites is very important, because it limits reactor development to the coastal areas, and in some coastal regions there is already significant generation overcapacity. Overcapacity in some parts of the coast has meant that reactors are being used for load-following—which A) they are not built for, and B) has a negative impact on their capacity factors and thus on their profit-

30. See, for example, World Nuclear Association, "Nuclear Power in Romania," last updated October, 2017, available from *http://www.world-nuclear.org/information-library/country-profiles/countries-o-s/romania.aspx* and Stephen Stapczynski and Aibing Guo, "China Widens Nuclear Exports With $7.7 Billion Romania Plant," *Bloomberg*, November 9, 2017, available from *https://www.bloomberg.com/news/articles/2015-11-10/china-keeps-nuclear-exports-going-with-7-7-billion-romania-deal*.

ability—and new plants have even been delayed in entering service because there is insufficient electricity demand, further affecting the economics of nuclear investments.[31]

Given the challenges above, it is unclear to the authors how long the China government will put its weight behind its nuclear export industry, when the nation as a whole could win much more business and gain more political influence by putting its weight behind other technologies such as high-speed rail, in which it has had remarkable domestic success, and renewable energy, in which it has substantial and sometimes dominant market shares both at home and abroad.

Nuclear Fuel Cycle Facilities

China obtains portions, in some cases most, of its uranium, fuel conversion, enrichment, and fuel fabrication needs (the "front end" of the fuel cycle) from domestic resources and facilities. As a nuclear weapons state, some of these facilities were originally developed in support of China's nuclear weapons program. The World Nuclear Association describes China's policy on uranium acquisition for its nuclear power program as targeting "about one-third of uranium supply domestically, one-third from Chinese equity in foreign mines, and one-third on the open market."[32] China has seven operating uranium mines, two of which date from the 1960s and 70s, as well as uranium resources in many other locations. China has purchased a significant amount of uranium from a number of supplier nations, and two Chinese firms hold equity in uranium mines in central Asia, Africa, and Canada.

---

31. See C.F. Yu (2016), "CHINA: Beijing Drafts Rule to Put Reactors In Load-Following Mode," *Nuclear Intelligence Weekly* 10, no. 33, August 19, 2016, available from *http://www.energyintel.com/pages/about_uiw.aspx*.

32. World Nuclear Association, "China's Nuclear Fuel Cycle," updated July 2017, and available from *http://www.world-nuclear.org/information-library/country-profiles/countries-a-f/china-nuclear-fuel-cycle.aspx*.

Enrichment of uranium is the process of concentrating the fraction of the U-235 isotope found in natural uranium (about 0.7%) to the fraction needed in LWR fuel, typically 3% to 5%. A preliminary step to enrichment is conversion of natural uranium oxide, a solid, to gaseous uranium hexafluoride (UF_6). The World Nuclear Association lists China's conversion capacity as somewhat uncertain, with a report of a 5000 tons per year t/yr plant in Gansu province operating at 80% of capacity, with another 9000 t/yr plant due to come on line this year or next. A smaller plant (500 tons of uranium metal per year (tU/yr)) plant is reported in northwest Gansu province. Another plant with a capacity of 3000 t/yr is reportedly being built by China Nuclear Fuel Corp in Hunan province and will be on line in 2018.

Enrichment of uranium in China takes place at larger plants in Shanxi and Gansu provinces, and at two smaller plants in Sichuan province. The total capacity of these plants is estimated at 5.7 to 7.0 million SWU (separative work units, a measure of enrichment capacity) per year in 2015, and a projected range of 10.7 to 12.0 million SWU/yr in 2020. By way of comparison, annual enrichment demand for Chinese reactors was expected to total 9 million SWU in 2020. Enriched uranium has also been imported to China, particularly for reactors of foreign design, with enrichment services provided in Europe and Russia. Centrifuges used for enrichment were provided by Russia in the past, but indigenous centrifuges technology has been used for recent capacity expansions, starting in 2010.[33]

The fabrication of fuel for Chinese reactors is done mostly in China, with some reactors supplied by France and Russia under contract to receive fuel from those nations sufficient for the first reactor core loading and a number of subsequent loadings. A plant in Yibin, Sichuan province, with a total capacity of 900 tU/yr, makes fuel for PWRs and for the Russian VVER design. A second major plant

33. Hui Zhang, *China's Uranium Enrichment Capacity: Rapid Expansion to Meet Commercial Needs*, Project on Managing the Atom, Discussion Paper No. 2015-03, Belfer Center for Science and International Affairs, Harvard Kennedy School, August, 2015, available from *http://www.belfercenter.org/sites/default/files/legacy/files/chinasuraniumenrichmentcapacity.pdf*.

in Inner Mongolia, at Baotou, fabricates fuel for China's pair of CANDU reactors as well as for various PWR models,[34] and will reportedly have total annual capacity of 1600 tU/yr by 2020. A facility at the Baotou plant is also making fuel for the high-temperature gas reactor being completed in Shandong province. Fuel pellets for the fuel assemblies for some reactor models are also sourced from Kazakhstan's Ulba Metallurgical Plant.

Most spent PWR fuel in China is currently stored in pools on reactor sites. Generally, the on-site spent fuel storage capacity at operational nuclear power plants can accommodate 10 years of spent fuel. Taking into account ongoing trends in nuclear fuel management, such as increasingly high rates of fuel burnup (which reduce the number of refueling cycles necessary), extensions of reload cycles, and the use of dense-pack storage in spent fuel pools into consideration, it is estimated that the storage capacity of present facilities can be enlarged to hold approximately 20 years' worth of spent fuel. Currently, all PWR spent fuel is in fact stored at reactor sites except for some of the spent fuel removed from the Daya Bay reactors, the first units in commercial operation in China. Since 2003, shipments of spent fuel from Daya Bay have been transported approximately twice annually to the centralized interim storage facility in Gansu province, where it is placed in wet storage facilities (away-from-reactor spent fuel pools).[35]

The facility in Gansu province, the Jiuquan Atomic Energy Complex (JAEC) as noted above, was initially developed in the 1950s and 60s to support China's nuclear weapons program. Since then, facilities for storage of civilian nuclear spent fuel have been added, as well as a pilot-scale reprocessing plant, constructed starting in

34. Including the AP1000 and its Chinese variant the CAP1000, based on Westinghouse designs and intended to be the "main basis of China's move to Generation III technology" (see World Nuclear Association 2017a, ibid).

35. Liu Xuegang, *Spent Nuclear Fuel Management in China*, Nautilus Institute NAPSNet Special Report, August 5, 2014, available from *https://nautilus.org/napsnet/napsnet-special-reports/spent-nuclear-fuel-management-in-china/*.

1986 but not operational until 2010, that can handle 60 tons of spent nuclear fuel (expressed as the mass of heavy metal—mostly uranium plus plutonium—abbreviated as tHM) annually.[36]

Plans and Projections for Reactor Deployment

The World Nuclear Association lists a total of 20 reactors in China as "under construction" and 30 to 40 more as "planned," mostly with an expected construction start before 2019. The under construction and planned units total 66.7 GWe.[37] In addition, the same reference lists 100 units, totaling 114.2 GWe, as "proposed", and an additional 79 units, and 90.8 GWe, proposed for the more distant future. Together, these listings plus the reactors already in operation total well over 300 GWe, over three times the size of the United States reactor fleet at its historic maximum in 2012.[38]

One recent article, citing both the World Nuclear Association source referenced above and a Chinese-language study by the Chinese Academy of Engineering lists ranges of nuclear capacities of 83.8 to 200 GWe by 2030, and 250 to 400 GWe by 2050.[39] Other re-

---

36. See, for example, Nuclear Threat Initiative, "Jiuquan Atomic Energy Complex," last updated September 29, 2011, available from *http://www.nti.org/learn/facilities/722/*.

37. World Nuclear Association, "Nuclear Power in China." For nuclear power plants, capacity is sometimes shown as GWe, or gigawatts of electric power, to differentiate from output expressed in terms of thermal power, which is typically about three times higher.

38. Nuclear Energy Institute, "US Nuclear Generating Statistics, 1971 – 2016", updated April, 2017, available from *https://www.nei.org/Knowledge-Center/Nuclear-Statistics/US-Nuclear-Power-Plants/US-Nuclear-Generating-Statistics*.

39. Qiang Yue, Jingke He, Laurence Stamford, and Adisa Azapagic, "Nuclear Power in China: An Analysis of the Current and Near-Future Uranium Flows," *Energy Technology* 5, no. 5, 2017 pp. 681–691, available from *http://onlinelibrary.wiley.com/doi/10.1002/ente.201600444/epdf*.

ported projections provide ranges of 340 to 500 GWe by 2050,[40] 1200 TWh of generation by 2040,[41] and on the order of 250-300 GWe, by 2050, the latter in a "two-degree" scenario where fossil fuel use is largely phased out except with the use of carbon capture and storage technologies.[42] These projections for massive growth in nuclear capacity are tempered by reports identifying the significant challenges to the Chinese nuclear sector in the years to come,[43] and focusing on renewable power and efficiency as alternatives to both fossil- and nuclear generation.[44]

---

40. Brian Wang, "China will complete five nuclear reactors in 2017 and double nuclear power generation to about 420 TWh by 2021," *NextBigFuture*, March 2, 2017, available from *https://www.nextbigfuture.com/2017/03/china-will-complete-five-nuclear.html*.

41. U.S. Department of Energy, Energy Information Administration, "China expected to account for more than half of world growth in nuclear power through 2040," September 28, 2016, available from *https://www.eia.gov/todayinenergy/detail.php?id=28132*. 1200 TWh at an average annual capacity factor of 80% would translate into about 170 GWe of nuclear generation capacity.

42. International Energy Agency/Nuclear Energy Agency, *Technology Roadmaps: Nuclear Energy*, 2015 Edition, available from *https://www.iea.org/media/freepublications/technologyroadmaps/TechnologyRoadmapNuclearEnergy.pdf*.

43. See, for example, Xu Yi-chong, "China's contested nuclear future: The expansion of China's nuclear power production faces some serious challenges," *Asia Pacific Policy Forum*, February 5, 2016, available from *https://www.policyforum.net/chinas-contested-nuclear-future/*. This article cites challenges to nuclear build-out including difficulties in reactor siting, the use of multiple technologies as an impediment to cost reduction and regulation, other challenges in organizing and making effective the regulation of the industry, and competitive realities associated with the slowdown in the growth of electricity demand coupled with the growth of deployment of and ongoing cost reduction in the renewable power industries.

44. Energy Research Institute, Lawrence Berkeley National Laboratory, and Rocky Mountain Institute, *Reinventing Fire: China*, Executive Summary, September 2016, available from *https://www.rmi.org/wp-content/uploads/2017/05/OCS_Report_ReinventingFireChina_2016.pdf*. Graphics in this document suggest that under a "Reinventing Fire" scenario, China's electricity generation from nuclear power in 2050 would be about 1300-1400 TWh, or the equivalent

Plans for Future Nuclear Fuel Cycle Facilities

The World Nuclear Association has summarized China's known plant for future development of "front-end" (nuclear fuel supply) and "back-end" (spent fuel management) fuel cycle facilities.[45] China looks to expand a number of mining activities, including, interestingly, extraction of uranium from coal ash from a power station in Yunnan province. Even with such expansions, based on at least one analysis,[46] it seems unlikely that China will be able to produce one-third of its uranium needs domestically, thus the significant emphasis by Chinese firms on investments in and joint ventures with companies prospecting for and producing uranium abroad.

Hui Zhang of Harvard University's Project on Managing the Atom, in a 2015 report, suggests, based on interviews with Chinese experts, that Chinese capacity to expand its enrichment capacity in the late 2010s was about 1 million SWU per year.[47] Given existing capacity, ongoing expansion, and the importation of initial cores for several reactors built with imported technology, Zhang estimates that China will easily have sufficient enrichment capacity to meet domestic needs through 2020, and perhaps will be able to sell surplus enrichment services internationally, "consistent with the government's stated policy of 'self-sufficiency' and 'targeting the international markets' in the supply of enrichment services." Information on China's plans for expansion of enrichment capacity beyond 2020 was not immediately available, but it seems likely that continued expansion of domestic facilities, coupled with the availability of capacity internationally, will be sufficient to fuel China's reactors on an ongoing basis.

---

of about 190 GWe of capacity at an average capacity factor of 80%.

45. World Nuclear Association, "Government Structure and Ownership: Nuclear Power in China Appendix 1."

46. Qiang Yue, Jingke He, Laurence Stamford, and Adisa Azapagic, ibid.

47. Zhang, 2015, ibid.

Data provided by the World Nuclear Association suggests that China will have fuel fabrication facilities for PWRs of about 2400 tU/yr by 2020, sufficient to meet demand by that year. Some fabricated fuel is imported, particularly, as noted above, for the new cores of reactors built with imported technology. Information on plans for expansion of fuel fabrication facilities, as with enrichment, was not immediately available, but it is assumed that CNNC, as the entity responsible for fuel fabrication in China, will continue to expand capacity to meet domestic demand, and possibly some fuel exports as well.

Reports indicate at least two ongoing efforts by China to build new facilities for reprocessing spent PWR fuel to separate plutonium for use in mixed-oxide fuel and, eventually, for use in fast reactors. A "medium scale" facility with a capacity of 200 tHM/yr is to be built at a site in Gansu province about 100 km away from the existing pilot-scale reprocessing plant described above. The medium-scale plant is part of an overall project of CNNC called "Long Teng 2020" (Dragon Soars 2020). The overall project was reported to have an expected cost 100 billion RMB (about USD 16 billion),[48] of which the reprocessing plant is an unspecified fraction, although an independent estimate provides a range of USD 3.2 to 5.7 billion (capital costs only).[49] In addition, CNNC is also negotiating with the multinational nuclear fuel cycle vendor Areva to develop an 800 tHM/yr reprocessing facility at a site on China's east coast. At 800 tHM/yr, this facility would be the size of Japan's Rokkasho reprocessing facility.

---

48. Hui Zhang, "China is said to be building a demonstration commercial reprocessing plant," *International Panel on Fissile Materials*, September 23, 2015, available from *http://fissilematerials.org/blog/2015/09/china_is_said_to_be_build.html*.

49. Matthew Bunn, Hui Zhang, and Li Kang, *The Cost of Reprocessing in China*, Project on Managing the Atom Report, Belfer Center for Science and International Affairs, Harvard Kennedy School, January 2016, available from *http://www.belfercenter.org/sites/default/files/legacy/files/The%20Cost%20of%20Reprocessing.pdf*.

Overall, as reported by the World Nuclear Association, China is researching (in a program dating back to 1964) and plans use of "recycled" fuel on three tracks, as summarized in Figure 6. First, mixed-oxide fuel (MOx), which blends uranium with plutonium recovered from spent fuel during reprocessing, will be used in existing and new PWRs. Second, MOx will ultimately be used for fuel a new set of Fast Neutron, or Fast Breeder Reactors (FBRs in the Figure below). A small test fast reactor, the Chinese Experimental Fast Reactor, was built near Beijing was with Russian collaboration, and started operation in 2011. A larger test fast reactor (600 MWe) is scheduled to be constructed stating in December 2017, with operation in 2023. Assuming a positive decision for deployment in 2020, China's first commercial fast reactor (the Chinese Commercial Fast Reactor, listed at 1000 to 1200 MWe) will be constructed starting in 2028, with commissioning in 2034. Fast reactor capacity of 40 GWe is "envisaged" by 2050. Third, a blend of plutonium and recycled uranium from reprocessing of PWR fuel, plus and depleted uranium from enrichment, will be used in pressurized heavy water reactors (CANDU units, or PHWRs). A test of the use of blended recycled fuels in unit 1 of the Qinshan Phase III PHWRs was carried out, and deployment of the recycled fuels in both PHWR units was planned for 2018.

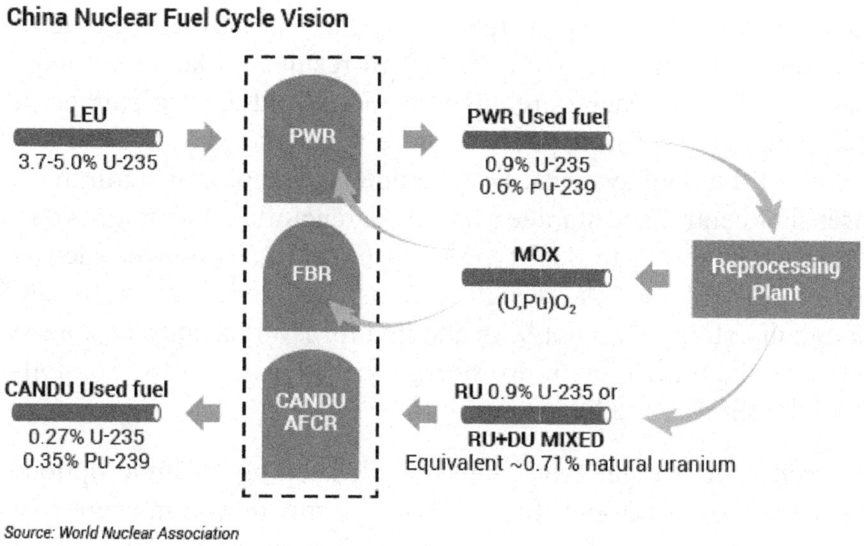

Figure 6. Nuclear Fuel Cycles Planned for China.[50]

Apart from reprocessing, with regard to spent fuel management in China, a report on spent fuel storage by Professor Liu Xuegang concluded, in part:[51]

"Dry storage is currently only used for CANDU reactors in China, and will be implemented for HTR spent fuel. These two reactor models account for only a minor portion of the whole Chinese nuclear fleet. But the utilization of dry storage and its performance will have a great impact on future decision-making for the sector. Though some experts consider that the pool capacity at reactors in China is large enough to accommodate spent fuel for the next 5 to 10 years, there are strong voices supporting the building of a large-scale centralized spent fuel storage facility soon. In part, it is argued, the current practice of pool storage for spent fuel in highly dense packed arrays has been subject to criticism follow-

50. World Nuclear Association, "Government Structure and Ownership: Nuclear Power in China Appendix 1."

51. Liu Xuegang, *Spent Nuclear Fuel Management in China.*

ing the Fukushima accident. In case dense-racking is ultimately not chosen as a means of spent fuel storage in China, the decrease in potential spent fuel storage density will result in a lack of storage space at Chinese reactors relatively soon. For at-reactor storage, it is difficult to build new pools to store spent fuel due to the complexity of the pool systems. Dry storage is very promising in those cases. For centralized storage away from reactors, dry storage is still a strong competitor to pool-type storage due to advantages such as low investment, modular design, and easy maintenance. As a result, though dry storage has not been adopted for PWR spent fuel storage in China, the utilization of dry storage facilities is a strong possibility in the short or medium-term."

In essence, it is clear that China is still keeping multiple options open, from once-through fuel cycles with medium-term spent fuel storage in dry casks to fast reactor options with reprocessing of spent PWR fuel, and authorities have not yet converged on a single path for the nuclear sector. But as spent fuel builds up at existing reactors, reaching decisions regarding the future management of nuclear spent fuel will become more imperative.

Summary of Business-As-Usual/Baseline/Reference Electricity Sector Scenario for China

Introduction

The current status of and recent trends in China's electricity sector, as described above, forms the basis from which the sector, and the nuclear energy components thereof, will evolve in the future. Although many types of "disruptive" events can occur to suddenly change how the electricity system may evolve—an accident at (as in Fukushima) or attack on a nuclear facility being a prime, but hardly exclusive, example—in general projections of electricity futures assume continuations of existing trends, informed by expected changes in demographics, economic development, regional, national, and international policies, and other "drivers." To systematically compare policy-driven scenarios for electricity sector development, it is important to start with a "reference" case for development of the sector (and the economy in which it is embedded) to provide a consistent basis for comparison of both qualitative and quantitative attributes between scenarios. Such a reference, or "business as usual" (BAU) case is described briefly below, prepared as a composite of several studies of China's electricity future. This BAU case serves as the basis for exploration of the alternative scenarios presented in Section 3 of this paper.

Description of Composite BAU Scenario

The BAU (or reference) scenario presented here represents a composite of a number of literature sources, including work by the U.S. DOE EIA, British Petroleum, Lawrence Berkeley National Laboratory, and others as described in the next section. The focus is on the future of electricity generation in China, and in particular, nuclear energy's role in same, but more general economy-wide metrics are provided as points of reference and comparison with other studies.

Some of the key general parameters of the BAU Scenario for China are as follows:

- Reflecting the maturing Chinese economy, gross domestic product (GDP) growth slows over time, averaging 5.8 percent annually from 2015 through 2020 and 5.4%/yrfrom 2020 through 2030, but slowing to 3.5%/yr from 2030 through 2050, and 2.5%/yr from 2040 through 2050.[52]

---

52. Projections shown here are consistent with those included in the U.S. DOE Energy Information Administration, "International Energy Outlook, 2017," as included in the downloaded table "World gross domestic product (GDP) by region expressed in market exchange rates," available from *https://www.eia.gov/outlooks/aeo/data/browser/#/?id=4-IEO2016&cases=Reference&sourcekey=0*. Note that GDP growth values are not used as direct drivers of electricity

- As projected by the United Nations, China's population growth slows through 2030, when it peaks at about 1.45 billion, declining thereafter to about 1.37 billion by 2050.[53]

- Growth in electricity generation also slows, continuing recent trends. Generation (and thus consumption, which is equal to generation less transmission and distribution losses) growth falls from 6.8% annually in the first half of the 2010s to 3.5%/yr from 2015 through 2020, then to 1.75%/yr from 2020 through 2030, 1.3%/yr from 2030 through 2040, and 0.8 percent annually thereafter through 2050.[54] This trend is another reflection of a maturing economy and, coupled with continuing (though declining) GDP growth, results in falling energy intensity per unit of GDP. The declining population after 2030, however, means that growth in electricity production (and use) per person grows faster than overall electricity use, increasing, for example, at 1.5%/yr from 2030 through 2040, and 1.2%/yr from 2040 through 2050.

- While total electricity generation nearly doubles, from about 5700 TWh in 2015 to over 10,000 TWh by 2050, shares of electricity generation in China continue to shift over time, as shown in Figure 7. The share of electricity output provided by coal-fired power declines from about 70% in 2015 to slightly over 40% by 2050, with growth in gas-fired, wind, and nuclear generation providing nearly all

consumption in the BAU scenario presented here, but are provided for reference.

53. United Nations Department of Economic and Social Affairs, Population Division, *World Population Prospects: The 2017 Revision, dated June, 2017.* Data shown are based on data for China in table "Medium fertility variant, 2015 – 2100." Downloaded File "WPP2017_POP_F01_1_TOTAL_POPULATION_BOTH_SEXES.xlsx."

54. These increases in electricity use are generally consistent with the BAU scenario for China included the U.S. DOE EIA *International Energy Outlook, 2017* document, as referenced above.

of the displacement. Hydroelectric generation, as a share of the total, changes relatively little over time, solar grows substantially but to only 4.1% of generation by 2050, and the use of liquid petroleum products, never a large fraction of generation in China, continues to decline. Nuclear power's share of generation grows to 13% by 2050, but most of the growth in its share of the power market occurs before 2030.

- Changes in generation capacity by type, as shown in Figure 8, reflect the same general trends as in electricity generation itself, except that wind and solar power, due to their lower capacity factors (operating fewer hours per year due the intermittent availability of wind and solar resources) relative to coal-fired and nuclear units, account for a larger share of added and total generation capacity. Natural-gas-fired power undergoes a shift from being mostly a peak resource, with low capacity factors (23% in 2015 and 2020) to being mostly a baseload resource (capacity factor of over 60% in 2050), reflecting a shift from simple-cycle gas turbines to more efficient combined-cycle plants.
- Total Nuclear generation capacity in the BAU case rises to about 140 GW in 2040, and 170 GW in 2050. As such, it assumes a slightly reduced growth trend, particularly after 2040, than the BAU projections for capacity provided in the USDOE's most recent International Energy Outlook, which calls for 139 GW of generation in 2040, and 187 GW in 2050.

Additional details on the nuclear fuel cycle elements of the BAU scenario are provided in the following section.

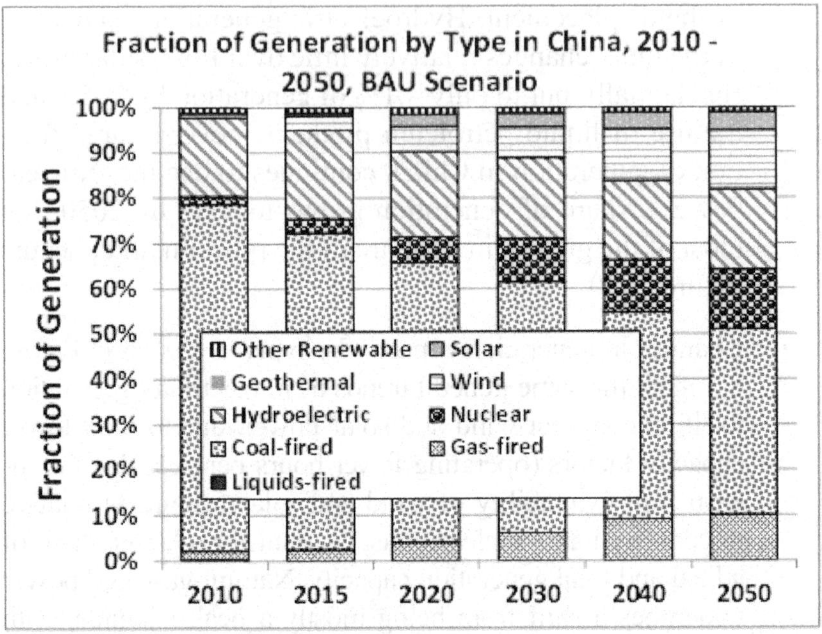

Figure 6. Fraction of Generation By Type in China, 2010-2050, BAU Scenario.

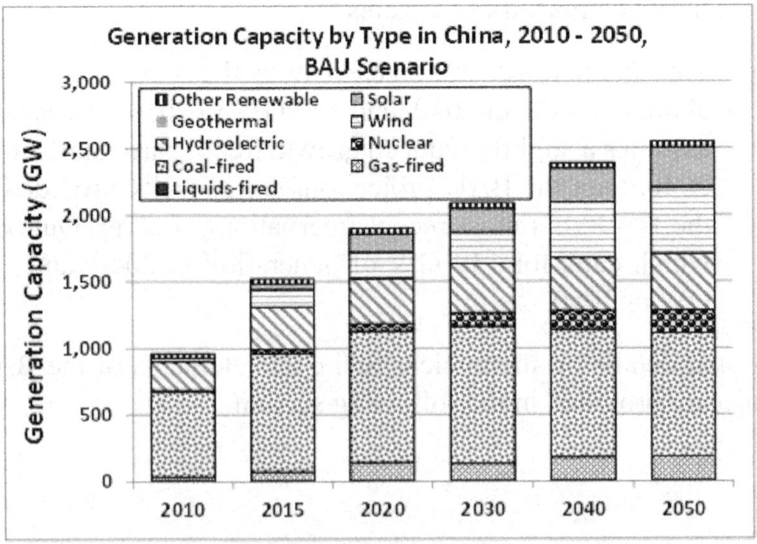

Figure 7. Generation Capacity By Type in China, 2010-2050, BAU Scenario.

Alternative Scenarios for China's Energy and Nuclear Future

Introduction—Focus of Alternative Scenarios and Key Examples

Alternative scenarios, or alternative projections (which are not necessarily the same thing), can be prepared to demonstrate different potential energy futures for a nation, state, province, city, region, or other jurisdiction. Scenarios and projections are thus tools that analysts, and the policymakers they serve or seek to influence, use to test how policies devised and implemented in the near or more distant future are likely to affect key metrics that are of concern to policymakers and to society as a whole. These metrics will typically include quantitative measures such as total energy or electricity use, total cost, and emissions of local and greenhouse gas pollutants, but may also include qualitative metrics such as the expected impact on energy or military security, environmental or political risk, or social impacts.

As there are literally an infinite number of future scenarios and projections that can be chosen, it is important for analysts to select examples for comparison that are plausible—though looking out more than 30 years into the future is by definition an exercise in speculation—yet are sufficiently different that the comparison of the different cases yield policy-relevant insights. At the same time, the different scenarios/projections should typically be configured to provide the same *energy services*, that is, for example, to support economies of roughly the same size and growth rate, to move the same number of people the same distance (if not always by the same modes), and to heat homes and cool to the same degree, though not always with the same energy sources. For the different China energy futures we describe below—focusing on electricity—we have defined scenarios/projections that support approximately the same economic structure and GDP growth rates, and the same populations, but are quite different in terms of how the energy systems, and especially the electricity and nuclear energy sectors, might evolve.

Given the global importance of the Chinese energy sector, a number of different groups have, over the past several years, prepared their own versions of how energy supply and demand in China might evolve through 2040 or 2050. Some notable examples include:

- The U.S. Department of Energy's Energy Information Administration projections as a part of their *International Energy Outlook* series, the 2016 and 2017 versions of which have been used to help define the BAU case projections for the Chinese electricity sector presented in the previous section of this paper.
- Work by the China Energy Group of Lawrence Berkeley National Laboratory (LBNL), including studies developing and analyzing scenarios for the evolution of China's energy sector through 2050.
- Work by the Rocky Mountain Institute and LBNL under the "Reinventing Fire" project, which looks at scenarios to vastly reduce energy sector greenhouse gas emissions in several countries, including China.
- A study of scenarios of the Chinese power sector by a group from the Renewable and Appropriate Energy Laboratory of the University of California-Berkeley.
- Work by Greenpeace, including on the future of nuclear and coal-fired power in China, and scenarios of accelerated deployment of renewable energy systems.
- Scenarios of the evolution of the energy sector published by the Energy Research Institute (ERI) of the National Development and Reform Commission (NDRC).
- Continuing work under the China Renewable Energy Outlook project, implemented by ERI and the China National Renewable Energy Centre (CNREC), with cooperation from the Danish Energy Agency and the U.S. National Renewable Energy Laboratory (NREL).

Brief summaries of each of these efforts are presented briefly below, followed by quantitative and qualitative comparisons of three "composite" scenarios that draw from these studies and others, and feature deployment of nuclear energy at different levels.

Summary of Examples of Alternative Scenarios

Alternative scenarios explore and project futures for China's energy sector that are different than a reference or business-as-usual case, and thus reflect the application of policies designed to steer the energy sector in a given direction, the influence of changes in the energy sector (or broader economy and society) not anticipated in the reference case, or both. Summaries of several alternative energy and, especially, electricity-sector scenarios for China are provided below. Most of these scenarios focus primarily on accelerated implementation of energy efficiency and/or renewable energy, relative to a reference case, but each could or would have significant implications for the nuclear energy sector as well.

Lawrence Berkeley National Laboratory (LBNL) China Energy Group

The researchers in the China Energy Group of the Energy Analysis and Environmental Impacts Division of LBNL have carried out a number of studies in which they have developed scenarios of China's energy future, usually focusing on energy efficiency and/or reducing greenhouse gas emissions.[55] A recent (2017) publication

55. See, for example, Nina Khanna, David Fridley, Nan Zhou, Nihan Karali, Jingjing Zhang, and Wei Feng, "China's Trajectories beyond Efficiency: CO_2 Implications of Maximizing Electrification and Renewable Resources through 2050," May, 2017, prepared for the European Council for an Energy Efficient Economy's 2017 Summer Study, available from *https://china.lbl.gov/sites/default/files/1-142-17_zheng_khanna_final.pdf.* The China Energy Group uses the LEAP (Long-range Energy Alternatives Planning) software tool developed by the Stockholm Environment Institute—United States to quantitatively model

by the China Energy Group compares a "reference case" with alternative cases for electricity demand. A "Cost-effective Efficiency and Renewables Scenario" reduced year-2050 electricity demand by 21%, relative to the reference case, and a "Maximum Electrification Scenario"—based on the Cost-effective Efficiency and Renewables Scenario, but incorporating additional electrification in all sectors—still reduced 2050 electricity demand by 13%. Both of these scenarios resulted in a reduction of over 50% in national CO_2 emissions in year 2050 relative to the reference case (from 11.57 billion tonnes, or Gt, CO_2 to 4.79 and 4.72 Gt CO_2, respectively). A third alternative scenario, including maximum deployment of demand-side renewables, offered additional reductions, to 3.98 Gt CO_2 by 2050. By way of comparison, China's 2016 CO_2 emissions stood at about 10 Gt, so each of the alternative scenarios represents a significant reduction, by 2050, relative to current emissions.[56]

In a 2016 study focused on energy demand and CO_2 emissions in China's cities—which produce on the order of 75-80% of China's national CO_2 emissions in recent years—LBNL China Energy Group researchers evaluated a "Low Carbon" scenario representing "a pathway in which commercially-available cost-effective efficiency and renewable energy technologies are fully deployed."[57] The measures included in the Low Carbon scenario reduced year

energy flows, emissions, and other parameters for China in LBNL's China 2050 Demand Resources Energy Analysis Model (DREAM).

56. See, for example, Zhu Liu, *China's Carbon Emissions Report 2016: Regional Carbon Emissions and the Implication for China's Low Carbon Development*, Environment & Natural Resources Program, Belfer Center for Science and International Affairs, Harvard Kennedy School, October, 2016, available from *https://www.belfercenter.org/sites/default/files/files/publication/China%20Carbon%20Emissions%202016%20final%20web.pdf* and David Fridley, Hongyou Lu, and Xu Liu, *Key China Energy Statistics 2016*, available from *https://china.lbl.gov/sites/default/files/ced-9-2017-final.pdf*.

57. Nina Z. Khanna, David Fridley, Lynn Price, Nan Zhou, and Stephanie Ohshita, "Estimating China's Urban Energy Demand and CO_2 Emissions: A Bottom-up Modeling Perspective," 2016 ACEEE Summer Study on Energy Efficiency in Buildings, available from *https://china.lbl.gov/sites/default/files/11_129.pdf*.

2050 reference case urban emissions by approximately two-thirds, a result consistent with the findings in the 2017 study detailed above.

In a 2016 study focusing on the implications of power-sector policies on coal-fired generation and CO_2 emissions in China, the China Energy Group projected that a "Strengthened renewable MMS [mandatory market share] with green dispatch scenario" could reduce coal-fired generation to 14% of total generation by 2050.[58] In the same scenario, renewable power generation (biomass, wind, solar, and hydro) provides about 63% of generation in 2050, with output of "renewables" (inferred in this case to mean wind, solar, and biomass) reported at 4472 TWh, and nuclear power providing about 20% of total generation. At that level of nuclear generation, the implied capacity of nuclear power assumed in the Strengthened renewable MMS with green dispatch scenario would be nearly 300 GW.[59]

Rocky Mountain Institute, LBNL, and Energy Research Institute "Reinventing Fire" Project

An ongoing collaboration between the Rocky Mountain Institute (RMI), LBNL, and the Energy Research Institute (ERI) of the of the National Development and Reform Commission of China (see below), the "Reinventing Fire" project seeks to identify pathways of deep carbon dioxide emissions reduction through a combination of energy efficiency, electrification of end-uses that currently use fossil fuels, and expanded deployment of renewable electricity.

58. Nina Zheng Khanna, Nan Zhou, David Fridley, and Jing Ke, "Quantifying the potential impacts of China's power-sector policies on coal input and CO_2 emissions through 2050: A bottom-up perspective," *Utilities Policy* 41, August 2016, pp. 128-138, available from *https://eta.lbl.gov/publications/quantifying-potential-impacts-chinas*.

59. Estimate of implied nuclear capacity by the author of this paper is based on review of Figure 3 in Khanna et al, 2016 (ibid), and assuming an average capacity factor for nuclear generation in 2050 of 85%. Reference case generation in 2050 was reported as 10,730 TWh.

The collaboration draws on work by all three partners, including the LBNL and ERI described above and below, respectively. An Executive Summary of the *Reinventing Fire* report from 2016 shows a reference case in which electricity generation rises to 12,800 TWh by 2050, up from 4,700 TWh in a 2011 base year.[60] In the "Reinventing Fire" case, electricity generation in 2050 falls by 16% relative to the reference scenario, to 10,800 TWh, as a result of two partially offsetting sets of policies. Energy efficiency reduces 2050 electricity demand by on the order of 6,000 TWh, relative to the reference case, while electrification displacing fossil fuel use throughout the economy adds back 4,000 TWh to the 2050 total. Additional renewable electricity generation in the "Reinventing Fire" scenario displaces about 3,000 TWh of fossil and some nuclear generation in 2050, relative to the reference case. Nuclear generation in 2050 in the "Reinventing Fire" case is about 1,550 TWh, as opposed to about 2,400 TWh in the reference case. These values correspond to 2050 nuclear capacities of about 210 and 320 GW, respectively, assuming an average capacity factor for nuclear generation of 85 percent in 2050.[61]

Real GDP growth in both cases was assumed to be slightly higher than that included in the BAU case outlined previously, above, at annual averages of 7.18% from 2010-2020, 5.60% from 2020-2030, 4.12% from 2030-2040, and 2.94% from 2040-2050. Population growth assumptions were similar to those in the United Nations "Medium Variant" projections reported above.

The net result of the "Reinventing Fire" scenario is to essentially cut in half year 2050 primary energy requirements relative to the reference case, bringing overall primary energy use by the Chinese economy back to 2050 levels. Overall 2050 CO_2 emissions similarly

---

60. ERI, LBNL, and RMI, *Reinventing Fire: China, A Roadmap For China's Revolution in Energy Consumption and Production To 2050*, Executive Summary, September, 2016, available from *https://www.rmi.org/insights/reports/reinventing-fire-china/*.

61. Nuclear generation values estimated by the author of this paper from data presented in Figure ES-14 in ERI, LBNL, and RMI, ibid.

fall by nearly half in the "Reinventing Fire" case, relative to the Reference case. Emissions of the local and regional air pollutants sulfur and nitrogen oxides fall by factors of approximately 8 and 12, respectively, relative to 2010 level, in the Reinventing Fire scenario. Further, the authors of the report find a net direct economic benefit due to the transition to the "Reinventing Fire" scenario. The "Reinventing Fire" Scenario will

> "...save China 21 trillion RMB in energy costs. From 2010 to 2050, implementing the Reinventing Fire Scenario yields a potential energy savings of 56 trillion RMB ($8.3 trillion) relative to the Reference Scenario. Incremental new investment required beyond the Reference Scenario to realize these energy savings is estimated to be 35 trillion RMB ($5.2 trillion), yielding a net present value savings of 21 trillion RMB ($3.1 trillion, all figures 2010 real)."

"SWITCH-China" Modeling by UC Berkeley Renewable and Appropriate Energy Laboratory

SWITCH-China is an integrated model of the Chinese electricity sector prepared by researchers at the University of California's Renewable and Appropriate Energy Laboratory (RAEL) and their colleagues.[62] In a 2016 study, the SWITCH-China model was used to investigate four scenarios of the evolution of carbon emissions by the Chinese power sector. The most stringent of these, the "IPCC Scenario [Intergovernmental Panel on Climate Change]," includes on the order of 250-300 GWe of nuclear generation capacity by

62. Gang He, Anne-Perrine Avrin, James H. Nelson, Josiah Johnston, Ana Mileva, Jianwei Tian, and Daniel M. Kammen, "SWITCH-China: A Systems Approach to Decarbonizing China's Power System," *Environmental Science and Technology* 50, no. 11, 2016, pp 5467–5473, available from *http://pubs.acs.org/doi/abs/10.1021/acs.est.6b01345*.

2050,[63] meeting 14% of total electricity demand, along with about 1500 GWe of wind power, about 1900 GWe of solar power, and 500 GWe of electricity storage. By 2050, about 90% of coal-fired generation (providing in total 29% of China's 2050 electricity needs) is coupled with carbon capture and sequestration (CCS) systems. CCS systems collect carbon dioxide from power plant exhaust gases (using considerable energy in the process) for sequestration in (typically) underground strata.

The SWITCH-China IPCC case results in a reduction of power-sector CO_2 emissions to 80% of 1990 levels. The authors of the study report that the annual additional cost of the scenario is over $2 trillion per year by 2050, which is offset at least partially ("22 to 42%") by the avoided external costs of coal production and use in the IPCC scenario. Placed in context, $2 trillion is about 5-7% of projected Chinese GDP in 2050. It should be noted that the SWITCH-China Reference Case appears to call for considerably higher growth in overall generation capacity, and thus likely in generation, relative to the composite BAU scenario described earlier in this appendix.

Commentary on Plans for Nuclear and Coal-fired Power Development by Greenpeace

Although the author of this paper was unable to find long-range projections of the evolution of the Chinese energy sector authored by Greenpeace that are similar to the studies described above, a number of Greenpeace publications have commented on the future of the Chinese coal and nuclear sectors. A 2012 article suggests that China should focus on renewable energy and energy efficiency to reduce greenhouse gas emissions, and that

> "...building enough nuclear power stations to make a meaningful reduction in greenhouse gas emissions

63. Capacity estimates are by the author of this paper based on review of figures in He et al.

> would cost trillions of dollars, create tens of thousands of tons of lethal high-level radioactive waste, contribute to further proliferation of nuclear weapons materials, and result in a Chernobyl-scale accident once every decade. Perhaps most significantly, it will squander the resources necessary to implement meaningful climate change solutions."[64]

A late 2017 article notes that China has recently canceled or delayed work on more than 150 planned or under-construction coal-fired power plants due to the "flat-lining" of demand for coal-fired power due to overcapacity of coal-fired generation. The article cites the environmental benefits of not moving forward with coal-fired power, including air pollution and water consumption benefits, and indicates that newly-increased targets for deployment of solar and wind power sources will further reduce the need for coal-fired generation. The article also notes that "[t]he government's recent efforts to clamp down on the red-hot real estate sector and local government debt spending—key drivers of China's heavy industry volumes and power demand—will also leave less space for coal-fired power generation."[65]

A Greenpeace report evaluated the co-benefits of renewable generation in China from 2015 through 2030.[66]

Among other findings, the report:

---

64. Greenpeace East Asia, "No nuclear in China - not anywhere," October 31, 2012, available from *http://www.greenpeace.org/eastasia/news/blog/no-nuclear-in-china-not-anywhere/blog/42752/*.

65. Lauri Myllyvirta and Li Danqing, "China halts more than 150 coal-fired power plants," October 11, 2017, available from *https://unearthed.greenpeace.org/2017/10/11/china-halts-150-coal-fired-power-plants/*.

66. Greenpeace, *Accelerating the Energy Transition The Co-benefits of Wind and Solar PV Power in China*, Summary, April 2017, available from *https://secured-static.greenpeace.org/eastasia/PageFiles/299371/Renewables%20co-benefits%20report,%20April%2017/Accelerating%20the%20Energy%20Transition_GPEA%20media%20briefing_0411.pdf*.

- Finds that the external environmental benefits of renewable generation exceed the wind power subsidy provided even in 2016,[67] and will be more than twice the level of the subsidy by 2030.
- States that "Between 2015 and 2030, wind and solar PV power will contribute RMB 14.3 trillion to China's GDP."
- Estimates that the wind and solar industries will employ 7.7 million people by 2030.
- Indicates that solar energy has been a force in "energy poverty" alleviation by being a major tool for providing electricity to households that previously lacked grid access.
- Notes that wind and solar development has reduced water use by coal-fired power plants, with the reduction by 2030 "... expected to increase to 3.6 billion m^3, equivalent to the annual basic water consumption of 200 million people."

A 2016 article provided by Greenpeace identifies some of the issues affecting deployment of reactors in China based on imported designs, and suggests that these issues will make China's ambitious nuclear power development targets difficult to achieve.[68]

China National Renewable Energy Center's "China Renewable Energy Outlook 2016" (CREO)

The report *China Renewable Energy Outlook 2016*, prepared by the China Renewable Energy Center in collaboration with groups in China, Europe, and the United States, focuses on an alternative

67. The subsidy location cited is in Zhangjiakou, Hebei Province, where the wind power generation subsidy was 0.14 RMB/kWh in 2016.

68. Zachary Davies Boren, "All is not well with nuclear in China," February 17, 2016, available from *http://energydesk.greenpeace.org/2016/02/17/not-even-china-likes-the-hinkley-epr-nuclear-reactor/*.

"High Renewable Energy (RE) Penetration" scenario to a "Stated Policies" reference case, both evaluated through 2030.[69] The two scenarios differ only modestly with regard to overall energy consumption, with the stated policies case reaching a slightly higher peak in overall energy consumption a few years later than the high RE penetration case, and having a somewhat slower decline in contribution of coal to primary energy use (47% of total primary energy in 2030 in the Stated Policy scenario versus 42% in the High RE Penetration scenario, down from 65% in 2016). The High RE Penetration case, however, provides a third more RE energy production by 2030 than the Stated Policies case. This difference is accomplished mainly by adding over 1000 GW of mainly wind and solar generation capacity in the High RE Penetration scenario, over and above the amounts assumed for the Stated Policies case. Coal-fired capacity falls from a high of 960 GW in 2020 in both scenarios, to 710 GW in the Stated Policies case and 660 GW in the High RE Penetration scenario, Generation from coal-fired power plants falls more dramatically, however, and as a result CO_2 emissions are 1.25 billion tonnes lower (about 13%) in the High RE Penetration scenario, and local air pollution and its health effects fall much more rapidly as well. The CREO report estimated a small increase in GDP in the High RE Penetration case relative to the Stated Policies case, with a significant boost to employment in renewable energy-related occupations partially offset by a reduction in employment in the fossil fuels sectors, mostly coal mining.

With regard to the nuclear sector, the CREO report includes the following description:

> "According to the decisions by the previous as well as current government of China, construction of nuclear power plants in the inland and in large-scale

---

69. China National Renewable Energy Center and Energy Research Institute (CNREC and ERI, 2016), *China Renewable Energy Outlook 2016*, prepared with cooperation partners the Danish Energy Agency and the U.S. National Renewable Energy Laboratory, available from *http://www.ea-energianalyse.dk/reports/1473_REO2016.pdf*.

> construction in the Yangtze River Basin will not happen. The development of the western regions has priority for a "green mountains and clear water are as good as mountains of gold and silver", primarily based on renewable energy. Before the fourth generation of nuclear power technology is in commercial operation, it is assumed that China will not open the inland deployment of nuclear power. Based on this, we consider nuclear power development to be within the range of 100 GW in 2050, solely deployed in coastal areas."

Although nuclear deployment in the two scenarios considered in the CREO 2016 report does not different markedly, with nuclear development limited to 75 GW by 2030, CREO thus projects much less nuclear capacity than that implied by the roster of planned reactors presented by the World Nuclear Association (as described above), but still creates significant environmental benefits, relative to the Stated Policies case, through the High RE Penetration scenario.

The CREO project partners have been updating and expanded their analyses, and a "CREO 2017" report is forthcoming.

<u>Energy Research Institute's "China 2050 High Renewable Energy Penetration Scenario And Roadmap Study"</u>

The 2015 report *China 2050 High Renewable Energy Penetration Scenario And Roadmap Study*, prepared by the Energy Research Institute (ERI) of the National Development and Reform Commission (NDRC), and funded by the Energy Foundation, presents scenario for the evolution of the Chinese energy sector that, like the CREO and LBNL studies, includes significant additional electrification, relatively modest growth in nuclear generation capacity and use, and continuing strong growth in the deployment of renewable power sources.[70] Key findings of the report, as relayed in its Executive Summary, include the following:

- Total power generation in 2050 is higher than that in the BAU scenario presented earlier in this appendix, due principally to higher growth in generation in the early years of the projection, though growth in the later years (2040-2050) is somewhat lower in the ERI projection.
- The fraction of generation provided by coal-fired power plants falls from nearly 75% in 2011 (the base year of the ERI study) to less than 7% in 2050.
- The fraction of generation provided by solar and wind power rises to nearly 54% of the national total by 2050, with hydropower providing another 14%.
- Nuclear power generation capacity grows to 100 GW by 2050, at which point it supplies 4.3% of total generation. In the ERI projection, nuclear power's share of generation stays between 4.8% and 3.7% from 2018 through 2050.
- Carbon dioxide emissions fall to about 3 GT CO_2 by 2050.
- Emissions of the local and regional air pollutants nitrogen oxides and sulfur oxides (NOx and SOx) fall to approximately 1970 levels (NOx) and about half of 1970 levels (SOx), on the order of a tenth of peak emissions of both gases (reached in 2010 and 2005, respectively).
- The average costs of generating and delivering electricity in ERI's high renewables scenario are only very modestly higher than in ERI's reference case, ranging from 0.672 RMB/kWh in 2030 to 0.685 RMB/kWh (about 10.4 U.S.

---

70. Energy Research Institute (2015), Executive Summary, *China 2050 High Renewable Energy Penetration Scenario And Roadmap Study*, April 2015, available from *http://www.efchina.org/Attachments/Report/report-20150420/China-2050-High-Renewable-Energy-Penetration-Scenario-and-Roadmap-Study-Executive-Summary.pdf*.

cents/kWh at current exchange rates) in 2050 in the high renewables case, "while in [the] reference scenario the average cost between 2030-2050 will stay flat around RMB 0.67/kWh."[71]

- Under the high renewables scenario, about 10 million jobs are added in the renewable energy industries between 2015 and 2050.

- Relative to the reference case, ERI's macroeconomic analysis of the high renewables scenario indicates modest changes in key economic indicators in 2050. Government spending is lower, imports and exports higher, and residential consumption up, but all within the range of a 0.23 and 1.44% difference between cases. Price levels in the two cases are virtually the same in 2050.

Overall, ERI's results suggest that a shift to a high renewables penetration scenario can provide very significant environmental benefits at minimal overall costs to the economy—though of course some sectors will come out better than others—and without large growth in the deployment of nuclear power.

Quantitative and Qualitative Comparison of Alternative Scenarios with Baseline

The alternative scenarios of and commentary on China's near- and medium-term energy future presented above indicated that serious consideration is being given both inside and outside of China to policies to transform China's economy and energy sector towards a much lower-carbon, higher-renewables, and more efficient system than exists today. Some of the work described above (which certain-

71. Quote from ERI, ibid. Note that projecting costs two or three decades into the future is necessarily an uncertain exercise, thus the main point here is that the costs in the high renewables case appear minimally different from those in the reference case.

ly does not exhaust the universe of China energy futures studies) posit a relatively modest role for nuclear power in China's electricity sector by 2050, with little or no growth over time in nuclear's share of generation, while others suggest a more important role. In reviewing all of the projections above, and those presented in this paper, it is important to consider the breathtaking changes ongoing in the Chinese energy sector. Trends in just the past few years—marked reductions in deployment and use of coal-fired power, marked reductions in the rate of growth of electricity demand, and vast increases in wind and solar generation capacity (along with reductions in wind and solar costs) are shifting the baseline upon which scenarios are constructed. These trends may rapidly make readily achievable scenarios that look unachievable today.

Below we present, in the broader context of the overall electricity-sector scenarios described above and elsewhere in this chapter, three alternative scenarios of nuclear power sector development in China, and offer a brief analysis of their relative benefits and costs, both quantitative and qualitative.[72]

Summary of Nuclear Capacity Scenarios

In order to explore the consequences of different scenarios of nuclear power capacity development in China, we have created three different capacity expansion cases. These are as follows:

- A "*Business as Usual,*" or *BAU* case, that draws from a very recent listing of planned and proposed reactors in China prepared

72. The analysis of benefits and costs defines "energy security" in a broad sense, including traditional energy supply security and net direct costs, but also including environmental, social and political, and military-related security. For a discussion of an analytical framework used with this broader concept of energy security, see for example, David von Hippel, Tatsujiro Suzuki, James H. Williams, Timothy Savage, and Peter Hayes, "Energy Security and Sustainability in Northeast Asia," published in the "Asian Energy Security" Special Section of *Energy Policy* 39, no. 11, November, 2011, pp. 6719–6730.

by the World Nuclear Association (WNA),[73] but assumes that the phase-in of new "planned" reactors will be somewhat slower than in the WNA listing,[74] and that about 55 GWe of the reactors listed as "proposed" by WNA will ultimately be built by 2050 (out of a total of 200 GWe of projects listed). Based on standard operating lifetimes, Existing LWRs are retired when they reach 40 years of service, and China's two "CANDU" reactor units are retired after 30 years of service. By 2050, nuclear generation capacity in the BAU case reaches a level that is slightly lower than the nuclear capacity projected in the U.S. DOE EIA's *International Energy Outlook 2017* reference case, and also slightly lower than estimated 2050 nuclear capacity the *Reinventing Fire* China Reference case.

- A "*Maximum Nuclear*" or *MAX* case, which also draws from the WNA listing but assumes a faster phase-in than in the BAU case of under-construction and "planned" units, and also assumes that about 80% (160 GWe) of the reactors listed in the World Nuclear Association table as "proposed" (or replacements similar in total capacity) are ultimately built and are all phased in by 2050. Existing LWRs are retired when they reach 50 years of service, so 10 years of life extension is assumed. By 2050, nuclear generation capacity in the MAX case is lower than that implied in RMI's "Reinventing Fire" scenario.
- A "*Minimum Nuclear*" or *MIN* case, in which the reactors listed by the World Nuclear Association as under construction or planned are all ultimately phased in by 2038, though on a much slower schedule (for those plants for which an operational date

73. World Nuclear Association, "Nuclear Power in China," updated July 2017, available from *http://www.world-nuclear.org/information-library/country-profiles/countries-a-f/china-nuclear-power.aspx*.

74. Some authors see evidence that China's nuclear build-out is not proceeding as rapidly as initially planned. See, for example, Steve Thomas, "China's Nuclear Power Plans Melting Down," *The Diplomat*, October 29, 2016, available from *http://thediplomat.com/2016/10/chinas-nuclear-power-plans-melting-down/*.

is given) than listed. After 2038, no additional plants are built, and as in the BAU case, older plants are retired as they reach the end of their standard operating lives. This lower capacity scenario could arise due to a combination of factors, including minimal growth in electricity demand in the 2030s and beyond, increased price competition from renewable energy, the availability of new and cost-effective electricity storage technologies, and/or perhaps a social backlash against nuclear power. In the MIN case, nuclear capacity begins to slowly decline about 2050, reaching a level by 2050 that is somewhat below that of the ERI and CREO scenarios described previously.

Readers should note that none of these paths account explicitly for potential shocks to the Chinese nuclear power industry, and to Chinese society as a whole, that might arise from a serious accident in a Chinese nuclear power plant. Such an event could have potentially devastating consequences for large populations.[75] The timing of such an event, should it occur, is not knowable in advance, although there is an argument that it is statistically likely over the time frame of these paths, given historical rates of major accidents per year of reactor operation.[76]

Figure 8 and Figure 9, respectively, show the capacity and electricity output implied by each of the three nuclear scenarios above. In

---

75. The author of this paper prepared a rough estimate of the consequences of an incident involving (accident at or attack on) a reactor and spent fuel pool at the Ling'Ao nuclear plant in Guangdong province that suggested that "worst case" releases of Cesium-137 might result in exposures sufficient to cause hundreds of thousands of premature cancer deaths and almost certainly require the abandonment of one or several big cities, depending on the prevailing wind direction at the time of the incident (D. von Hippel and P. Hayes, 2016, unpublished).

76. See, for example, He Zuoxiu, "Chinese nuclear disaster "highly probable" by 2030," *China Dialogue*, March 19, 2013, available from *https://www.chinadialogue.net/article/show/single/en/5808* and S. Wheatley, B. Sovacool, D. Sornette, "Of Disasters and Dragon Kings: A Statistical Analysis of Nuclear Power Incidents and Accidents," *Risk Analysis* 37, no. 1, pp. 99-115, available from *http://onlinelibrary.wiley.com/doi/10.1111/risa.12587/epdf.*

the BAU case, capacity increases to 170 GWe by 2050, while in the MAX case capacity rises to nearly 260 GWe. In the MIN case, capacity rises to about 91 GWe by 2040 (about 260% of existing capacity as of 2017), but no new plants are added thereafter, so capacity falls to about 85 GWe by 2050 as older plants are retired. By way of comparison, as noted above, China's current (2016) overall electricity generation capacity for all types was over 1600 GW, and overall generation was nearly 6200 TWh.[77]

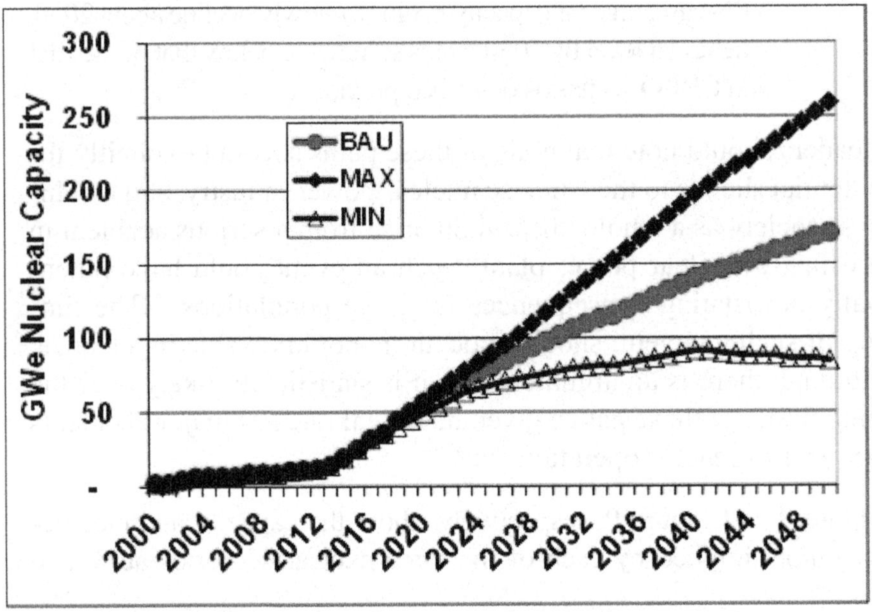

Figure 8. Three Scenarios of Nuclear Generation Capacity (GWe) for China through 2050.

77. For electrical energy, see British Petroleum, "BP Statistical Review of World Energy June 2017" workbook, available from *http://www.bp.com/content/dam/bp/en/corporate/excel/energy-economics/statistical-review-2017/bp-statistical-review-of-world-energy-2017-underpinning-data.xlsx*. For capacity, see, for example, "Installed capacity of electric power generation in China between 2010 and 2016 (in GW)", available from *https://www.statista.com/statistics/302269/china-installed-power-generation-capacity/*.

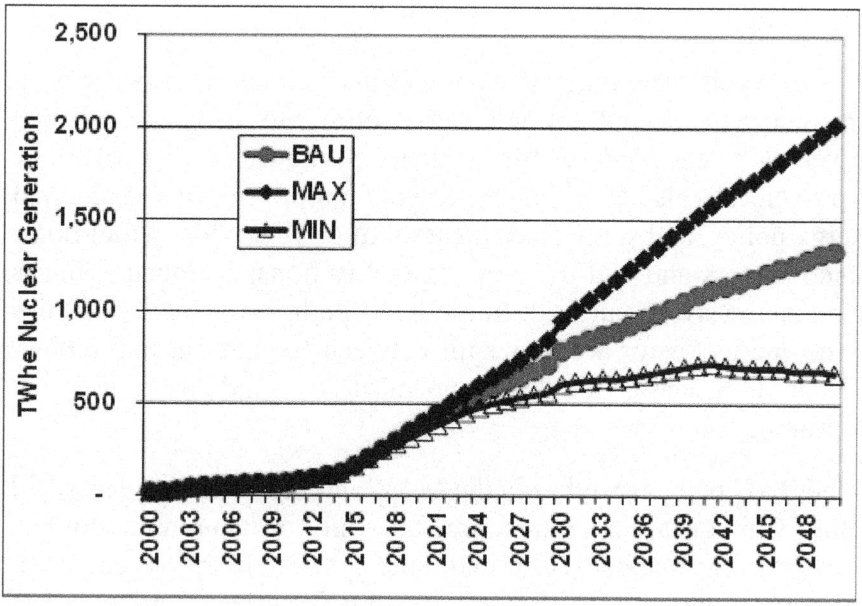

Figure 9. Three Scenarios of Nuclear Generation Output (TWhe) for China through 2050.

<u>Broader Energy and Nuclear Sectors Contexts for Nuclear Paths</u>

Each of the nuclear generation capacity expansion paths could, in theory, be combined with a range of different trends and policies in the broader energy and electricity sectors, as well as in the overall nuclear sector, including front-end and back-end fuel cycle developments. In practice, competition for private and public investment resources, national and societal goals and preferences, and other criteria make some combinations more plausible than others. The assumed energy and nuclear/sector contexts for each of the paths described above is as follows.

<u>BAU Path</u>

The BAU path is assumed to exist within a Chinese energy sector in which significant, but not aggressive, efforts to displace coal-fired power with renewable energy sources continue, energy efficiency improvements also continue, but are not major priorities of national energy policy, and some electrification of currently fossil-fuel-dominated sectors and end-uses, most notably transport, occurs, but is again aggressively pursued. In the BAU path, investment priorities in the energy sector are thus split between the nuclear and renewables sectors, with some development of coal- and gas-fired power continuing.

In the BAU path, as well as in the MAX and MIN paths as described below, China is assumed to source one-third of its uranium domestically.[78] In the nuclear sector under the BAU path, the development of uranium enrichment and reprocessing facilities, and the use of MOx fuel, proceeds roughly as described by the World Nuclear Association, but on a somewhat delayed schedule. For enrichment, this means that about 80% of China's enrichment needs are supplied domestically by 2020, with enrichment being entirely domestic by 2030. World (and Chinese) enrichment costs, driven in large part by Chinese nuclear expansion, follow a "medium" scenario, rising to about $75/kg SWU (about twice 2017 levels, but much lower than historical maxima) by 2050 (2009 dollars). Research and pilot development of fast reactor technologies continues, but also on a somewhat delayed schedule, such that commercialization of fast reactors is still at least several years off by 2050. MOx use is in LWRs assumed to start in 2025,[79] and 25% of reactors are assumed

---

78. From World Nuclear Association, "China's Nuclear Fuel Cycle," updated September 2017, available from *http://www.world-nuclear.org/information-library/country-profiles/countries-a-f/china-nuclear-fuel-cycle.aspx*. "China aims to produce one-third of its uranium domestically, obtain one-third through foreign equity in mines and joint ventures overseas, and to purchase one-third on the open market."

79. See, for example, Qiang Yue, Jingke He, Laurence Stamford, and Adisa Azapagic, "Nuclear Power in China: An Analysis of the Current and Near-Future

to use cores with 20% MOx fuel by 2050. Reprocessing also starts in 2025, with capacity and throughput ramping up sufficiently to process 50% of cooled spent fuel by 2040 and thereafter.

Spent fuel management in the BAU path, consistent with the path's emphasis on reprocessing, focuses mainly on interim spent fuel storage in spent fuel pools at reactors and at reprocessing facilities, although there is some dry cask storage of spent fuel.[80]

MAX Path

Consistent with, for example, the "Reinventing Fire" scenario described above, the MAX Path includes both significant investments in energy efficiency and in renewable generation, as well as in electricity transmission facilities. Electrification of the energy sector is also a priority, with the reduction of greenhouse gas emissions and local/regional air pollutants a driving policy impetus. Both the savings in year 2050 electricity needs due to energy efficiency improvements and additional electrification to displace fossil fuel at the end-use level are assumed to be about half that assumed in the "Reinventing Fire" study—about 25 and 17%, respectively. The lowered assumptions for these impacts are in part due to a lower reference case assumed here than assumed in the "Reinventing Fire" study, and in part to an assumption that a more aggressive build-out of the nuclear sector, and of, for example, reprocessing (see below) and enrichment, will to some

---

Uranium Flows," *Energy Technologies* 5, 2017, pp. 681 –691, available from *http://onlinelibrary.wiley.com/doi/10.1002/ente.201600444/epdf.*

80. "Dry casks" refers to the spent fuel storage option, used in many countries, in which cooled spent fuel is isolated in massive casks designed to last for up to 100 years. Spent fuel assemblies are typically sealed into stainless steel containers several centimeters thick, which are then placed in "overpacks" made of steel or concrete. The resulting cask is on the order of 2 meters in diameter and 6 meters tall, weighs up to 50 tonnes, and is essentially impervious to any significant damage that could be caused by natural disasters or accidents, as well as to all but the most determined and well-equipped attempts at penetration by criminals or terrorists.

extent crowd out investments in efficiency and renewable energy (for example). Investment to drive the nuclear sector comes to a large extent from public funding, while renewable power investments are mostly privately driven, but take advantage of necessarily higher prices for electricity required by nuclear investments.

The MAX path focuses on building national capacity in the nuclear sector, including aggressive build-outs of uranium enrichment and related capacity, as well as reprocessing capacity and MOx use. For enrichment, this means that about 90% of China's enrichment needs are supplied domestically by 2020, with enrichment being entirely domestic by 2025. Enrichment costs, again driven by Chinese demand, are assumed to follow a "high" scenario, rising to $104/kg SWU (still much lower than the $160/kg SWU historical maxima) by 2050. National capacity for manufacturing of key nuclear plant components grows rapidly, and the MAX path is likely to be also combined with an aggressive national effort to export nuclear reactors to other nations. MOx use is in LWRs assumed to start in 2022, and 40% of reactors are assumed to use cores with 20% MOx fuel by 2050. Reprocessing on a commercial scale likewise starts in 2022, with a target of 80% of cooled spent fuel reprocessed by 2035. Spent fuel management focuses on interim storage in spent fuel pools at reactors and at reprocessing facilities.

<u>MIN Path</u>

Consistent with paths projecting high rates of renewable energy penetration in the CREO and ERI studies, the MIN path couples nuclear capacity expansion at a rate that holds the nuclear share of generation roughly constant with aggressive development of renewable energy for power generation and for end-uses, plus aggressive energy efficiency programs. Electrification is also an emphasis, both for greenhouse gas emissions reduction and to bring down emissions of local and regional air pollutants. Energy efficiency efforts, and the use of more renewable energy sources at the end-use level, are assumed to reduce electricity needs by 33% by 2030 relative to BAU requirements, with electrification adding 20% back to electricity demand by 2050. Ongoing policies, including carbon markets and carbon taxes, preferentially target markets for renewable energy and energy efficiency, with the proceeds used to support investment in both, as well as in pollution control and environmental remediation. The MIN path does not explicitly include carbon capture and sequestration (CCS) for coal- and/or gas-fired power plants, as used in the "IPCC Target" scenario of the Gang He, et al (2016) study referenced above, but could include CCS if CCS technology is suitably advanced and if greenhouse gas emissions reduction policies in China required its use. China has been rapidly developing facilities for importing LNG, and LNG could play a more important role in power generation in China's future in a variant of a MIN path, with or without CCS. Greater gas-fired generation, perhaps displacing some planned nuclear plants, would be even more plausible if, for example, infrastructure for bringing North American gas to China at attractive prices can be developed. Also not explicitly modeled in the MIN path is the development and widespread use of electricity storage technologies, which would be needed to complement the aggressive development of renewable (wind and solar) electricity sources.

In the nuclear sector, reprocessing is not pursued beyond the existing and under-construction pilot plants, and MOx use in existing reactors is limited. Uranium enrichment facilities planned for the near-term are built by 2025, but no additional enrichment plants are built, and the remainder of China's required enrichment services are imported. With a smaller nuclear capacity expansion in China, and (assumed) reduced nuclear restarts in Japan, relative to other scenarios, international (and Chinese) enrichment costs follow a "low" trajectory, reaching $54/kg SWU by 2050, about 40% above 2017 levels. Efforts to export reactors continue, but are not heavily subsidized by the Chinese government. Research into fast reactors continues, but at a low level, and commercialization of fast reactors is still decades away by 2050. In part to provide fuel for fast reactor research, the smaller (200 tHM/yr) reprocessing facility reportedly (as of 2017) being built at Gansu is eventually completed, but is not run until 2025, and then at only partial capacity for several years. Plutonium (Pu) from this facility is used to produce MOx fuel, which is used in 10% of LWRs by 2050, with phase-in occurring slowly starting in 2030. MOx fuel again makes up 20% of reactor cores in those units that use MOx. Cooled spent fuel is stored mostly in a mixture of at-reactor spent fuel pools and in dry casks, which may be located at or near reactors or at a centralized dry cask facility.

Flows of Nuclear Materials

Table 2 shows estimated total requirements for uranium and uranium ore in each of the three scenarios, both for the individual years 2010, 2030, and 2050, and on a cumulative basis from 2015 through 2050. The MAX path implies the use of over 50,000 tonnes of U annually by 2050, while less than a third as much is required in the MIN path. The extraction of 11 million tonnes of ore is required by 2050 in the MAX path, but placed in context, this is much less than a percent of the total volume of coal extracted to fuel China's power sector. By assumption, two thirds of Chinese uranium needs are sourced abroad, and one third are from domestic mines.

Parameter	YEAR	MAX Path	BAU Path	MIN Path
Annual Total Metric Tons Natural Uranium (as U) Imported plus Domestic Production	2010	2,629	2,629	2,629
	2030	27,718	21,837	15,946
	2050	53,659	35,106	16,570
	Cumulative, 2015-2050	1,105,629	820,318	540,581
Annual Total Thousand Metric Tons Uranium Ore (from In-country and outside mines) to Supply All Domestic Uranium Needs	2010	546	546	546
	2030	5,752	4,532	3,309
	2050	11,136	7,286	3,439
	Cumulative, 2015-2050	229,457	170,245	112,190

Table 2. Uranium Requirements by Nuclear Capacity Expansion Path.

Under the MAX expansion path, if China chose to provide all of its own enriched uranium, China alone would need to build new enrichment capacity by 2050 approximately equal to more than half of today's global capacity. China's annual requirement requirements by 2050 rise to nearly 42 Million kg SWU. Under the MIN expansion path, however, international enrichment facilities extant as of 2015 are likely sufficient to meet China's enrichment needs by 2050 (about 12 million kg SWU), even factoring in likely East Asia regional and out-of-region demand without significant expansion, assuming existing international enrichment facilities (or replacement facilities) continue to operate. Though the ROK and Japan have accounted for almost all enriched uranium in East Asia needs pre-Fukushima, the rapid growth of China's nuclear power sector and the slow process of restarting Japan's reactors means that China's demand for enrichment will likely outstrip needs in the rest of the region well before 2020.

The uranium oxide (UOx) and MOx fuel requirements under each scenario are summarized in Table 2. MO fuel requirements in the BAU path by 2050 are less than half of those in the MAX path, and MIN path MOx use is less than a tenth of MAX path use by 2050.

Parameter	YEAR	MAX Path	BAU Path	MIN Path
Implied Total Requirements for UOx Fuel (excluding MOx) from All Sources (Metric tonnes heavy metal in fabricated fuel)	2010	267	267	267
	2030	2,929	2,327	1,711
	2050	5,417	3,620	1,763
	Cumulative, 2015-2050	115,040	86,601	57,961
Implied Requirements for MOx Fuel Blended and Fabricated from All Sources for Use in Domestic Reactors (Metric tonnes heavy metal in fabricated fuel)	2010	-	-	-
	2030	60	24	-
	2050	408	191	36
	Cumulative, 2015-2050	4,657	2,123	394

Table 3. Requirements for UOx and MOx Fuel in Each of three Nuclear Paths.

With widely varying use of reprocessing in the three scenarios, it is not surprising that plutonium production and uptake (as MOx) also is significantly different, as shown in Figure 10. Cumulative plutonium separation through reprocessing rises to over 450 tonnes of Pu in the MAX case by 2050, and about 9 tonnes of Pu inventory remains after MOx use in 2050. About 200 tonnes of Pu are produced by 2050 in the BAU case, but almost all stocks are used as MOx. In the MIN case, only about 27 tonnes of Pu are produced via reprocessing in China through 2050, and more than that amount of Pu is used as MOx, implying, for example, that some Pu from other nations (stocks from Russia, the UK, or France) is blended into fuel used in Chinese reactors, and/or military Pu from the Chinese weapons program is disposed of as MOx. Note that all of these calculations of net Pu inventories by 2050 are extremely sensitive to the combination of assumptions regarding the fraction of spent fuel reprocessed and the amount of fuel used as MOx. In practical terms, this means that if a reprocessing program is successful BUT MOx use is delayed, significant inventories of Pu can build up, and can serve as a proliferation target. As a sensitivity analysis, Figure 11 shows what might happen if reprocessing proceeds in each path as indicated above, but MOx use is delayed by 10 years in each path, and used in only half as many reactors—a plausible outcome given the difficulties in implementing MOx use that other nations have experienced to date. In this sensitivity case, inventories of Pu of about 300, 140, and 18 tonnes build up in the MAX, BAU, and MIN cases, respectively, by 2050, at which point Pu stocks are still continuing to accrue. The stocks building up in this sensitivity case in the MAX and BAU paths would suffice to build tens of thousands of nuclear weapons, and even the MIN case stocks represent thousands of times the mass of Pu contained in a nuclear warhead. Additionally, paths that produce high volumes of Pu, whether or not fully consumed as MOx, would tend to enhance the chance of significant volumes of Pu going astray, as the more Pu is produced, the easier it will be for weapons-relevant quantities (kilograms) of Pu, amounting to less than a tenth of a percent of MAX and BAU annual output from reprocessing facilities, to be diverted for criminal purposes.

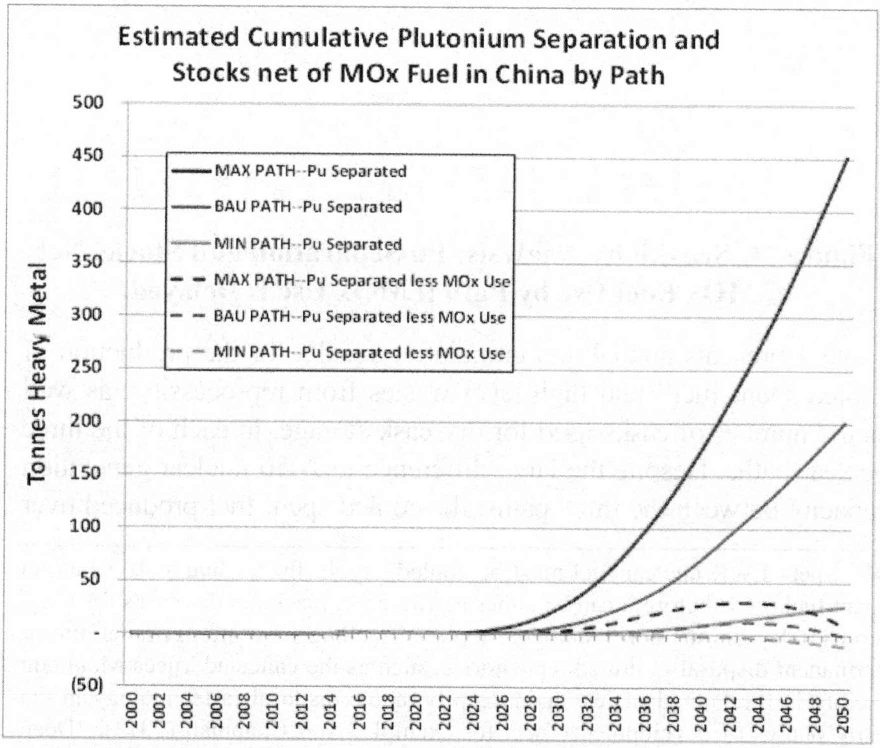

Figure 10. Pu Separation and Stocks Net of MOx Fuel Use by Path.

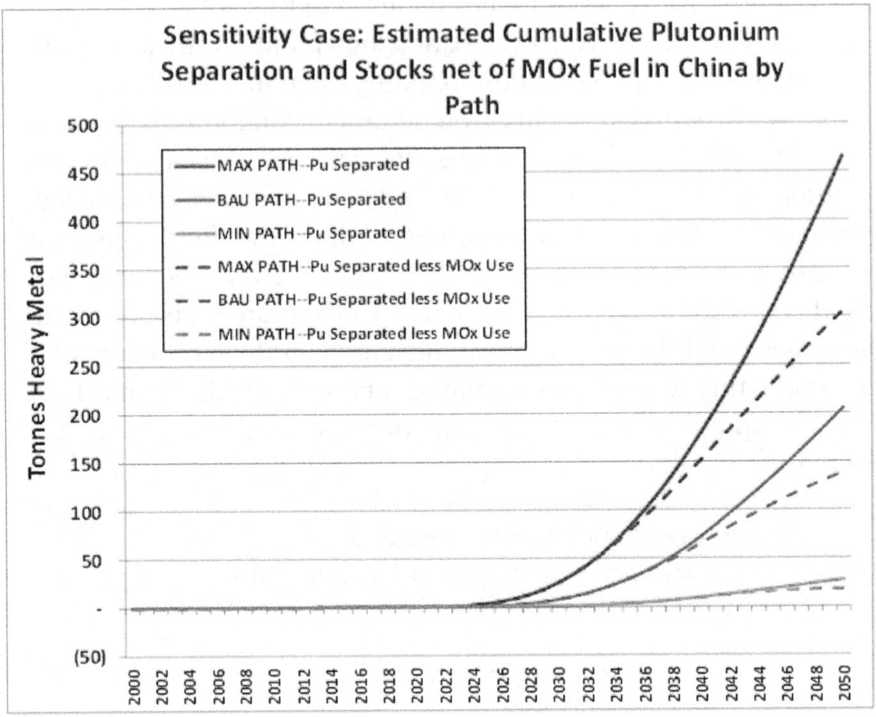

Figure 11. Sensitivity Analysis: Pu Separation and Stocks Net of MOx Fuel Use by Path if MOx Use is Delayed.

Table 4 presents annual and cumulative results for the production of cooled spent fuel[81] and high-level wastes from reprocessing, as well as the number of casks used for dry cask storage, in each of the three nuclear paths. Despite the large difference in 2050 nuclear generation capacity between the three paths, the cooled spent fuel produced over

81. Spent LWR nuclear fuel must be cooled—typically for five to 10 years, in spent fuel pools before it can be either reprocessed, placed in dry casks for long-term (100 years or more) storage, or placed in long term storage/indefinite or permanent disposal in mined repositories, such as the canceled Yucca Mountain facility in the United States, or in deep borehole disposal (a technology in the early phases of development—see, for example, Neil Chapman (2013), "Deep Borehole Disposal of Spent Fuel and Other Radioactive Wastes," *NAPSNet Special Reports*, July 25, 2013, available from *https://nautilus.org/napsnet/napsnet-special-reports/deep-borehole-disposal-of-spent-fuel-and-other-radioactive-wastes/*.

time is not all that different, because cooled spent fuel production lags changes in generation. The MAX scenario produces much more high-level waste (HLW) from reprocessing than the other cases. Though the volume of HLW is not large—5000 or so cubic meters could be contained in 25 or so average urban apartments—it is highly radioactive and, like spent fuel, remains so for thousands of years, meaning that special well-secured facilities capable of holding the wastes indefinitely must be constructed. In the MIN path, where the emphasis is on dry cask storage, on the order of 2600 casks would be required to accommodate the cooled spent fuel produced by 2050, not including the relatively small amount of spent fuel reprocessed in the MIN path. To put this number of casks in perspective, the total dry casks filled during the MIN path could be stored in an area of less than 10 hectares, just a bit bigger than area enclosed by the fence around the White House in Washington, DC.

Parameter	YEAR	MAX Path	BAU Path	MIN Path
Annual New Spent LWR Fuel Cooled and Available for Reprocessing, Storage, or Disposal (excluding MOx spent fuel), Metric Tonnes Heavy Metal	2010	51	51	51
	2030	1,190	1,139	975
	2050	3,654	2,686	1,747
	Cumulative, 2015-2050	57,028	46,688	35,880
Annual Spent MOx Fuel Cooled and Available for Storage or Disposal, Metric Tonnes Heavy Metal	2010	-	-	-
	2030	-	-	-
	2050	239	112	23
	Cumulative, 2015-2050	2,020	883	150
Implied Amount of Spent LWR Fuel (UOx only) Reprocessed in Total for Use in Domestic Reactors (Metric tonnes heavy metal)	2010	1	1	1
	2030	612	214	9
	2050	2,923	1,343	175
	Cumulative, 2015-2050	41,045	18,276	2,438
Implied Volume of High-level Waste (as vitrified) from All Cooled Spent LWR Fuel (UOx only) Reprocessed for Use in Domestic Reactors (cubic meters)	2010	0.15	0.15	0.15
	2030	70.37	24.57	1.02
	2050	336.18	154.47	20.10
	Cumulative, 2015-2050	4,720.15	2,101.75	280.35
Total Annual Number of Casks Required for Dry Cask Storage of Cooled Spent LWR Fuel (UOx only) for Storage/Disposal, Net of Reprocessing (units)	2010	-	-	-
	2030	-	-	-
	2050	36	109	156
	Cumulative, 2015-2050	362	1,496	2,608
Total Annual Number of Casks Required for Dry Cask Storage of Cooled Spent LWR MOx Fuel for Storage/Disposal (units)	2010	-	-	-
	2030	-	-	-
	2050	24	11	2
	Cumulative, 2015-2050	202	88	15

Table 4. Spent Fuel Management Results for Three Nuclear Paths.

<u>Costs</u>

Although a full quantitative evaluation of the direct costs of the scenarios described above is beyond the scope of this paper, some of the considerations in comparing the costs between the three scenarios in qualitative terms are as follows:

- Fossil fuel costs for electricity generation will be highest for the BAU case, with annual costs in the MAX case about half of BAU levels by 2050, and in the MAX case about a quarter of BAU levels, due to displacement of fossil fuels by efficiency improvements, electrification, and use of renewable energy for direct end-uses and electricity generation.

- Overall fossil fuel costs, including "upstream" costs for oil refining, coal mining, and other fuel-cycle activities, will be lower due to electrification of end-uses in the MAX scenario, and even lower in the MIN case.
- There is likely to be very little growth in capital and operating and maintenance (O&M) costs for fossil-fueled power plants in any of the scenarios, though these costs will be lower in the MAX path, and lower still in the MIN path, as more coal-fired generation is displaced by nuclear (in the MAX path) and renewable generation/energy efficiency in both paths. In the MAX and especially the MIN path, at least more coal-fired capacity will likely be retired, providing a reduction in O&M costs. One factor that could cause power plant costs to rise, and could be applied to any of the three scenarios, is if more stringent controls are required to reduce local and regional air pollutants, and/or if carbon capture and sequestration is required on fossil-fueled plants.
- Costs for energy efficiency and increased deployment of end-use renewables will be higher in the MAX case than in the BAU case, and higher still in the MIN case, although previous experience in China and elsewhere suggests that in many cases energy efficiency provides electricity savings more cheaply than electricity case be generated by supply-side resources. Thus when the avoided costs of electricity generation and direct fuel use are factored in, efficiency investments will likely yield a net savings in overall direct costs. The much higher rate of renewables deployment in the MIN path will likely reduce the per-unit costs of renewable energy systems, and accompanying electricity storage systems. It is difficult to say whether

the net costs of aggressive deployment of energy efficiency and renewable energy in the MIN path will be higher or lower than the costs of the conventional and nuclear energy systems that they displace in the BAU and MAX paths, but recent experience and at least some previous studies of energy futures suggest that the cost differences may be small relative to overall energy system costs and to the uncertainties of costs projected three decades into the future.

- The MAX scenario will have the highest overall costs for nuclear generation capacity, as well as for fixed and non-fuel variable operating and maintenance (O&M) costs, followed by the BAU case. MIN case total capacity costs and O&M will be substantially lower than in the other two cases. Costs for fast reactor research and development will be highest in the MAX case, and relatively limited in the MIN case.

Nuclear fuel cycle costs—exclusive of reactor capital and non-fuel O&M costs—have been quantified for each of the three nuclear paths described above. Not surprisingly, as shown in Figure 12, nuclear fuel cycle costs are much higher in the MAX path, at a about $560 billion on a cumulative basis over 2015 through 2950. This total is nearly twice as much as for the BAU path, and on the order of five times that of the MIN path. The three largest cost categories are raw uranium and uranium enrichment—higher in the MAX path due to higher cost escalation assumptions—and reprocessing costs, which are nearly avoided altogether in the MIN path.

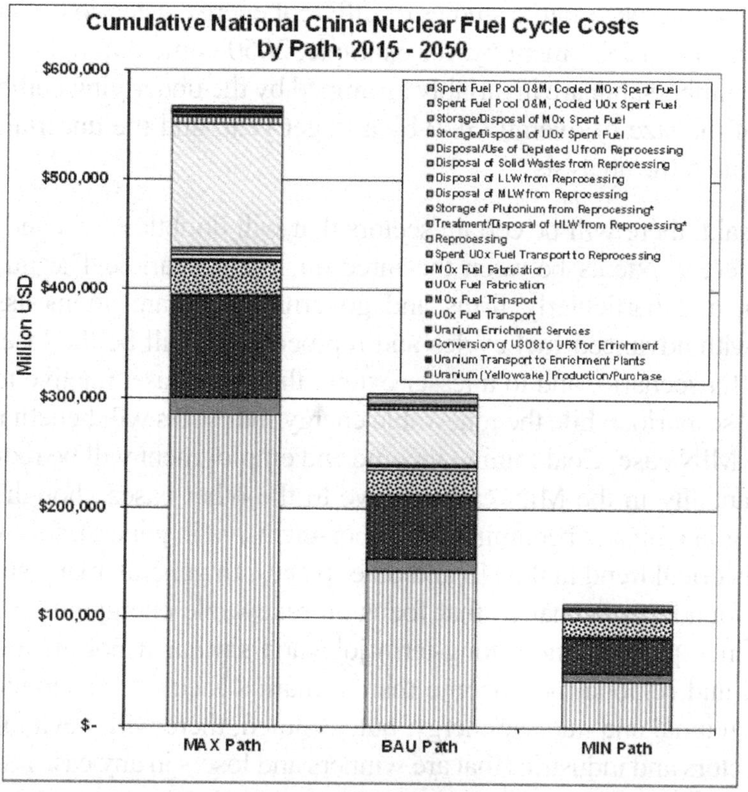

Figure 12. Cumulative National China Nuclear Fuel Cycle Costs by Path, 2015- 2050.

The relative indirect costs to the economy of alternative energy scenarios, and the policies that will drive them, are often a key consideration for policymakers. The perceived and projected impacts of different nuclear scenarios on, for example, GDP and employment at the provincial and national levels in China will have a considerable impact on the acceptability of particular scenarios and policies. In practice, there are always winners and losers—for example, with regard to employment in different sectors and even regions—when policies are shifted, but the net impact of these changes is very hard to know in advance, due to policy- and non-policy-related shifts in the economy and technology (and in underlying costs and factor prices) shifts over time, and to the general uncertainty associated with any economic prognostication.

In many macroeconomic studies of different energy scenarios, changes to GDP and employment by, for example, 2050 come out looking like large numbers, but are invariably swamped by the underlying combination of the size of overall GDP by a target year, and the uncertainties inherent in the analysis.

That said, there will be certain sectors that will doubtless win and lose to different extents between the three nuclear scenarios. The nuclear sector, and particularly firms and government organizations associated with advanced fuel cycles and reprocessing, will be the losers in the MIN scenario, and to a lesser extent, the BAU case, relative to the MAX scenario, while the renewable energy industries will benefit most in the MIN case. Coal mining income and employment will be reduced substantially in the MIN case relative to the other cases, though coal mining in China is becoming less labor-intensive in general, following the historical trend in the US and other places. In general, many studies have found that scenarios that focus on renewable energy and energy efficiency produce more long-term jobs and similar if not greater national and regional net income than scenarios focused on supplies of conventional and nuclear energy, but, as noted, there will inevitably be subsectors and industries that are winners and losers in any case relative to any other.

Energy Supply Security

The traditional concept of energy supply security, in brief, is that the more a country can source its fuel requirements from its own territory—or failing that, the more a country can draw for its energy needs from a diversity of domestic resources and imports from large number of trading partners—the more energy supply security is enhanced. Under that definition, the MAX case, which uses more nuclear power and less oil (the reduction being mostly due to electrification of the transport and other sectors) coal, and natural gas, is provides arguably more energy security than the BAU case. The MIN case, with more use of domestic resources (wind, solar, and energy efficiency) than the other two cases, arguably comes out on top in terms of energy supply security. Figure 13 shows the relative use of fossil fuels, both cumulative and for selected years, under each case. By assumption, in each case, uranium is sourced from the same ratio of domestic and foreign sources, though in principle, a higher proportion of domestic U could be used in the BAU and, especially, the MIN cases, given the lower overall U requirements.

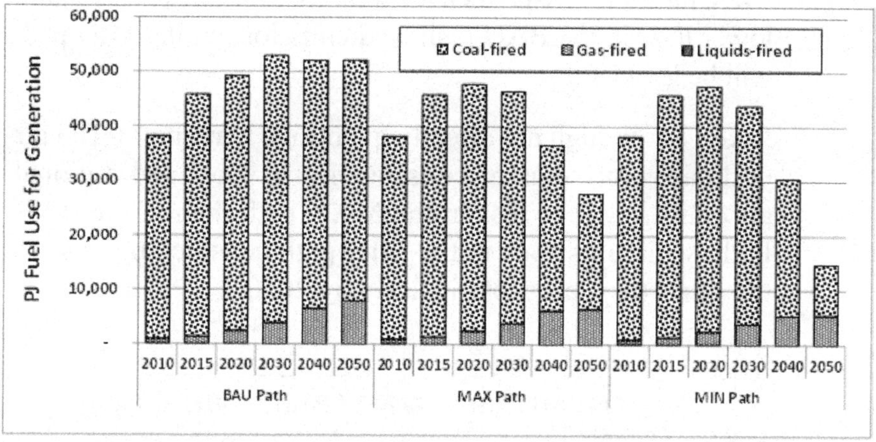

Figure 13. Annual Fossil Fuel Use for Electricity Generation by Type and Year for Three Paths.

<u>Environment</u>

The environmental component of a broader concept of "energy security" includes comparing the performance of the three scenarios on the basis of emissions of greenhouse gas emissions, local and regional air pollutant production, water pollution, and disposal of nuclear-fuel-cycle related wastes. With respect to these criteria, the three nuclear paths described above yield the following quantitative and qualitative comparisons:

- The MIN path produces nearly 30% lower cumulative electricity generation sector (2015-2050) greenhouse gas emissions than the other BAU path, and in particular, as shown in Figure 14, produces a small fraction of the year-2050 emissions included in the BAU and MAX paths. Although the overall Chinese societal GHG emissions are the main concern here, and have not been quantified for this paper, the inclusion of additional electrification in the MAX path indicates that China's MAX-path GHG emissions will be considerably lower than in the BAU path, and emissions in the MIN path will be lower still.
- Likewise, though not directly quantified here, non-GHG air pollutant emissions of consequence to local and regional air quality will be less in the MAX path than in the BAU Path, with emissions in the MIN path considerably less in the other two cases, particularly by 2050.

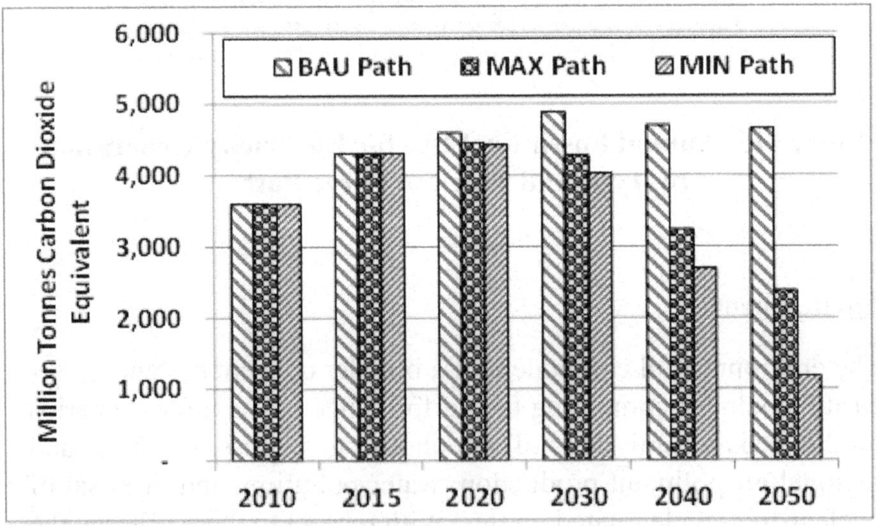

Figure 14. Greenhouse Gas Emissions from the Electricity Generation Sector in China Under Three Nuclear Power Development Paths.

- Because considerably less coal is used in the MIN and MAX paths than in the BAU path, water pollution from coal mines will be less in those paths. Thermal pollution from coal-fired power plants will also be less, though additional thermal pollution from nuclear power plants, some of which may be on inland sites (on rivers), may occur in the MAX path, relative to the other two paths. Additional water pollution and solid wastes from uranium mining in China and abroad will accrue in the BAU and, especially, MAX paths relative to the MIN path.
- The disposal of nuclear-sector-related waste streams will be much more of an issue in the BAU and, especially, MAX paths relative to the MIN path. Considerable HLW and intermediate and low-level wastes from reprocessing will accrue in the BAU and MAX paths, and China will need to find a final resting place for those materials, as well as for any spent fuel placed in long-term storage/disposal. Finding places to store/dispose of these materials may prove to be foci for political and social problems, as noted below.

Social and Political Criteria, and Military Security

During China's period of rapid economic growth, Chinese decisionmakers have typically given (or at least, exhibited) limited concern to the reaction of local populations in decisions on siting of key energy-sector facilities, relative to decisionmakers in many Western nations. Over the past decade, however, trends have suggested that the role of Chinese civil society in the siting of large and potentially polluting or dangerous facilities has been growing, at least in some ways. Very recent events have arguably suggested that civil society's voice in China may not continue to develop as some in the West might have hoped, though the impact of recent changes in Chinese governance on the nuclear sector is not yet clear.[82] In

82. See, for example, Chris Buckley and Adam Wu, "Ending Term Limits for China's Xi Is a Big Deal. Here's Why," *The New York Times*, March 10, 2018,

general, paths, like the MAX path, that call for large, centralized, secure facilities for handling and managing nuclear materials may galvanize opposition to such facilities on the local and national levels, thus making those paths arguably less secure than paths like the MIN path, where energy needs are supplied by resources that are often tapped by facilities that are more distributed and each smaller and less obtrusive (and polluting) than the other two paths.

Additionally, the nuclear facilities (including enrichment and reprocessing, nuclear power plants, and spent fuel transport) that are part of the BAU and, particularly, the MAX paths will require much more in the way of military security arrangements than the MIN path. These military security requirements increase military costs and enhance the possibility of conflicts between the military security apparatus and a population becoming accustom to greater social and economic freedoms. The impact of required nuclear sector security arrangements will be mirrored, to some extent, by the greater needs to secure supplies of oil and oil transport lanes; the needs to secure oil supplies will be highest in the BAU case.

Conclusions

What Alternative Scenarios of China's Nuclear Future Tell Us

As of 2017, China arguably sits at a point of decision, inflection, or possibly both in the evolution of its electricity generation system, its nuclear power future, and possibly its energy sector as a whole. Scenarios of nuclear power development in China, including those presented in this paper and many others, span the range from modest growth to 80-100 GWe by 2050 (from about 34 GWe today) to projecting growth to 300 to 400 or more GWe of nuclear power by 2050, with the beginnings of commercialization of fast reactor tech-

---

available from *https://www.nytimes.com/2018/03/10/world/asia/china-xi-jinping-term-limit-explainer.html*.

nologies. At the same time, transitions are occurring for China's coal-fired power fleet, with plans for future new capacity being rapidly scaled back, and smaller, less-efficient units being taken out of service. The rethinking of plans for expansion of coal-fired power is in part in response to progressively more stringent policies to address local and regional air pollution, but also in response to the two trends, particularly in recent years, of accelerating deployment of wind and solar power, and reduced growth in electricity demand, the latter particularly in comparison to the double-digit growth rates of recent decades.

Developing internally-consistent scenarios of China's electricity sector in general, and the nuclear energy sector within the electricity sector, with the different scenarios designed to serve the same needs for energy services in similar economic futures, provides a means to test policy directions. China could choose a path including rapid deployment of LWRs with spent fuel reprocessing, blending the resulting plutonium into MOx fuel and subsequent use of same in LWRs and, ultimately, in a fleet of fast reactors. Or an explicit or implicit policy (for example, through adjusting levels of power sector subsidies) could damp down the current nuclear build-out after those reactors currently under construction are built, such that additions in the decades after 2025 are modest, while aggressively encouraging energy efficiency and the development of solar and wind power, plus the supply-side changes (transmission systems, smart grids, and electricity storage, for example) that would be needed to maximize renewable energy usability. The estimates of future fuels use, costs, pollutant and waste emissions, and accompanying (typically) qualitative consideration of issues such as the relative political and social security ramifications that result from consideration of different future scenarios provide a way of testing and illustrating for policymakers the different ways of organizing the energy future of a nation.

Prospects for Meeting China's Future Energy Needs with Limited or No Increases in Nuclear Capacity and Proliferation-resistant Fuel Cycles

The comparison of the BAU, MAX, and MIN paths for the Chinese nuclear and electricity sectors, considered together with the existing body of China scenarios work described (in part) in this appendix, suggests that it will be possible for China to meet its economic development, GHG and air pollutant emissions reduction, and other goals without an extended and massive build-out of LWR capacity, and without expansions of uranium enrichment or reprocessing capacity beyond those projects now underway. Further, although the MIN case implies that China will not become a major exporter of nuclear power technologies, it also implies that China will continue along its current trend of being perhaps a dominant provider of renewable power systems. Nuclear sector costs in the MIN Path are much lower, both on an aggregate basis and per unit of output, than in the other two paths, largely because of lower uranium, enrichment, and reprocessing costs. Although these costs are only a small part of the overall cost of providing energy services to the Chinese economy, indications from past experience and other studies is that the emphasis on energy efficiency and renewable energy will offer the opportunity for China to effectively address its environmental concerns without significant (if any) additional costs, relative to a reference path. Further, a path with less nuclear power and fewer front-end and back-end nuclear facilities will be arguably easier to deploy in a social and political sense, particularly as expectations for a stronger voice in how its future unfolds continue to grow among the Chinese citizenry.

The MIN scenario provides significant benefits over the other two cases in terms of plutonium production and stocks (transient and otherwise), and thus provides significantly lower risk of the proliferation of nuclear weapons. For the MIN path to become a reality, policy support for energy efficiency and renewable energy will need to take precedence over policy support for nuclear power. Trends in recent years, including the slow-down in reactor construction and re-thinking of nuclear safety regulations post-Fukushima, ongoing structural change in the Chinese economy away from heavy industry (and much-reduced growth in electricity needs), and exceedance of even the ambitions government targets for renewable energy all point toward the enhanced practicality of a low-nuclear path for the evolution of the Chinese energy sector. Acknowledgement of the benefits of the MIN path (or similar) for China by international political and trading partners, probably including international policies that encourage such a path and embracing energy paths of their own that de-emphasize nuclear power, enrichment, and reprocessing, would likely serve to encourage China to move toward a low-nuclear future.

China arguably is at a point in marketing its nuclear technologies abroad in which its technologies are not particularly competitive—as they are based on older U.S. and other Western designs—and it is facing a worldwide market for nuclear power that even a Russian reactor vendor has reportedly described as weak. If China's domestic market for nuclear power were to follow a trajectory more like the MIN path described above than the BAU or MAX paths, it seems likely that China's nuclear exports would be relatively de-emphasized. Building and maintaining the capabilities to export nuclear technologies, including to countries where nuclear weapons proliferation is a danger (or historical fact, as in Pakistan), will be technically and economically riskier and more difficult without a burgeoning domestic market to fall back on. As such, timely encouragement (including by example) of China by the United States and the rest of the international community to focus on non-nuclear technologies for power generation could contribute to influences already in play and induce China to focus its efforts on exporting technologies that do not carry a weapons proliferation threat.

ABOUT THE CONTRIBUTORS

KYUNG-JIN BOO is a visiting professor in the Technology Management, Economics and Policy Program (TEMEP) at the College of Engineering at Seoul National University. Before he joined the faculty at TEMEP in March 2012, Dr. Boo was a researcher at Korea Energy Economic Institute (KEEI). He has been a Korean Representative for the ISO Strategic Advisory Group on Energy Efficiency and Renewable Energy Sources (SAGE) since 2008. Dr. Boo received his B.S. in Mining & Petroleum Engineering at Seoul National University in 1979 and his M.B.A. in International Business at the same university in 1983. Dr. Boo's Ph.D. studies were done at the University of Delaware in Public Economics and Policy.

RYOICHI KOMIYAMA is an associate professor at the University of Tokyo. He obtained a Ph.D. degree in 2003 from the University of Tokyo and joined the Institute of Energy Economics Japan (IEEJ) in 2003, working as a senior researcher. In 2010, he joined the faculty at the Department of Nuclear Engineering and Management, the University of Tokyo. In the meantime, he studied as a visiting researcher in Lawrence Berkeley National Laboratory from 2007 to 2009 and University of California at Berkeley from 2011 to 2012. His fields of research are energy modeling, economic analysis of nuclear energy and international energy market projection.

JOHN LEE is an Adjunct Professor and Michael Hintze Fellow for Energy Security at the Centre for International Security Studies, Sydney University, and a visiting scholar at the Hudson Institute in Washington, DC. Dr. Lee is the author of *Will China Fail?*, which was released by CIS in 2007 and the second edition in 2009. He is a prominent expert in Chinese development and the foreign policies of states in East and Southeast Asia as well as Indian security policy. He received his doctorate from University of Oxford.

KENNETH B. MEDLOCK III is the James A. Baker, III, and Susan G. Baker Fellow in Energy and Resource Economics at the Baker Institute and the senior director of the Center for Energy Studies. He is also the director of the Masters of Energy Economics program, holds adjunct professor appointments in the Department of Economics and the Department of Civil and Environmental Engineering, and is the chair of the faculty advisory board at the Energy and Environment Initiative at Rice University. Medlock is a principal in the development of the Rice World Natural Gas Trade Model, which is aimed at assessing the future of international natural gas trade. In 2012, Medlock received the prestigious Haydn Williams Fellowship at Curtin University in Perth, Australia. He is also an active member of the American Economic Association and is an academic member of the National Petroleum Council. Medlock has served as an advisor to the U.S. Department of Energy and the California Energy Commission in their respective energy modeling efforts. Medlock received his Ph.D. in economics from Rice University in May 2000.

JI-CHUL RYU is a retired Senior Fellow at the Korea Energy Economics Institute (KEEI). He is an expert in the areas of energy strategic development research, international energy cooperation, and energy planning and modeling. He joined the KEEI in April 1987. From 1996 to 2000 Dr. Ryu was the Vice-President of the Asia Pacific Energy Research Centre (APERC) in Tokyo, Japan. He also served as a Senior Counselor to the Minister of Energy and Resources at the Korean Government in 1991–1992. Dr. Ryu holds a Ph.D. in economics from the Australian National University and a B.Sc. in mathematics and statistics from the Seoul National University.

GU ZHONGMAO was a professor at the China Institute of Atomic Energy (CIAE), from 1970 to 1995. His work focused on research and development on radioactive waste treatment and disposal, including bituminization and cementation of lower and intermediate level waste (LILW), compaction of solid waste, separation of actinides from high level waste (HLW) and verification of HLW and Membrane separations, particularly liquid membranes. From 1996 to 2001, he served as the Director of the Radiochemistry Department, CIAE. From 2002 to 2010, he was the Vice-Chairman of the Science and Technology Committee, CIAE, where his research field covered back-end nuclear fuel cycles, including nuclear waste treatment and disposal, spent fuel reprocessing for nuclear power plants, and strategic studies on nuclear energy and the fuel cycle in China. From 2011 until the present day, Professor Gu has worked as a member of the Science and Technology Committee at CIAE.

HENRY D. SOKOLSKI is the Executive Director of the Nonproliferation Policy Education Center (NPEC). He previously served as Deputy for Nonproliferation Policy in the Department of Defense, and has worked in the Office of the Secretary of Defense's Office of Net Assessment, as a consultant to the National Intelligence Council, and as a member of the Central Intelligence Agency's Senior Advisory Group. In the U.S. Senate, Mr. Sokolski served as a special assistant on nuclear energy matters to Senator Gordon Humphrey (R-NH) and as a legislative military aide to Dan Quayle (R-IN). Mr. Sokolski is author of *Underestimated: Our Not So Peaceful Nuclear Future* (Strategic Studies Institute, 2017) and *Best of Intentions: America's Campaign Against Strategic Weapons Proliferation* (Praeger, 2001).

HIROSHI TAKASHI is a professor of Tsuru University. He specializes in energy policy and power market deregulation from a perspective of comparative policy studies. He was a member of the Electricity System Reform Committee and Special Adviser to the Cabinet Office of the Japanese Government. He is the author of the book, *Electric Power Liberalization*, published by Nikkei Publishing Inc.

DAVID VON HIPPEL is a Nautilus Institute Senior Associate based in Eugene, Oregon. His work with Nautilus has centered on energy and environmental issues in Asia, and particularly in Northeast Asia. He is currently involved in several Nautilus initiatives, including the multi-nation Asia Energy Security Project, centered around energy paths analysis, and the related East Asia Science and Security Network, which focuses on potential nuclear materials issues in the region. Also for Nautilus, he is involved in preparing an update to Nautilus' DPRK Energy Sector Analysis and in the DPRK Building Energy Efficiency Training project. In addition to his work with Nautilus, he works and has worked for a number of private and public agencies, including the World Bank, the United Nations and Tellus Institute/the Stockholm Environment Institute-US (Boston, MA), on a wide variety of topics ranging from utility planning and energy efficiency in the Middle East to reducing greenhouse gas emissions in several states and the Midwest region in the United States. Dr. Von Hippel holds M.S. and Ph.D. degrees in Energy and Resources from the University of California-Berkeley, and M.A. and B.S. degrees from the University of Oregon.

Made in the USA
Monee, IL
04 November 2026

67378675R10167